INS AND OUTS OF SCHOOL FACILITY MANAGEMENT

MORE THAN BRICKS AND MORTAR

Tak Cheung Chan
Michael D. Richardson

Published in partnership with the
Association of School Business Officials International

ScarecrowEducation
Lanham, Maryland • Toronto • Oxford
2005

Published in partnership with the
Association of School Business Officials International

Published in the United States of America
by ScarecrowEducation
An imprint of The Rowman & Littlefield Publishing Group, Inc.
4501 Forbes Boulevard, Suite 200, Lanham, Maryland 20706
www.scarecroweducation.com

PO Box 317
Oxford
OX2 9RU, UK

British Library Cataloguing in Publication Information Available

Library of Congress Cataloging-in-Publication Data

Chan, Tak Cheung, 1947–
  Ins and outs of school facility management : more than bricks and mortar / Tak Cheung Chan,
Michael D. Richardson.
    p. cm.
  "Published in partnership with the Association of School Business Officials International."
  Includes bibliographical references and index.
  ISBN 1-57886-191-8 (pbk. : alk. paper)
  1. School facilities—United States. 2. School plant management—United States. 3. School
buildings—United States.—Maintenance and repair. I. Richardson, Michael D. (Michael
Dwight), 1949– II. Association of School Business Officials International. III. Title.
LB3218.A1C53 2005
371.6—dc22
                                                                    2004017078

∞™ The paper used in this publication meets the minimum requirements of
American National Standard for Information Sciences—Permanence of
Paper for Printed Library Materials, ANSI/NISO Z39.48-1992.
Manufactured in the United States of America.

CONTENTS

pay close attention

PREFACE

When school districts plan to shrink their annual budgets, the school maintenance budget is always the first item on the chopping block because the negative effects of a diminished maintenance budget are not always apparent. But the damages to the school buildings are definite. A school building with reduced maintenance deteriorates quickly. The negative effects multiply themselves with an interrelationship between the building systems. Certainly, the consequences are serious enough to substantially shorten the useful life of a school building. The decision to cut school maintenance to balance a current school district budget is shortsighted. Little or no consideration is given to safeguarding the school district's long-term assets. Eventually school buildings have to be replaced sooner than anticipated or require a large sum of money for major repairs caused by delayed maintenance. The logic behind cutting a school maintenance budget is indefensible.

On the other hand, when school buildings are aging, attention is usually given to planning for replacement buildings. Extending the lives of old school buildings by renovation is seldom seriously considered. It is true that if a school building has no historical value, it is not worth keeping when renovation costs exceed 50 percent of the cost of new construction. However, most school buildings, except for those with structural deficiency, can be attractively renovated at reasonable cost to serve their new educational function. If more school buildings are planned for long years of extended use, then fewer new school buildings need to be planned. The result is a substantial savings of resources that can be reserved for other educational program development.

It is disheartening to see that many school districts fail to properly maintain their school facilities or to take advantage of the full potential of their existing school buildings. These school districts then turn around and ask the community to support the planning of new school buildings. The administrators of these school districts are accountable to the community, and they owe the community an explanation of what has been done to make use of public resources in protecting and developing existing school facilities.

The purpose of this book is to tell school administrators, school board members, and community members that existing school buildings need to be well maintained to be functional and that after renovation many old school buildings can continue to serve useful educational purposes for years. In economically tight years, planning for effective school maintenance and renovation programs makes sense in the long run.

Special thanks for the completion of this book are expressed to Dr. Theodore Kowalski, Dr. Ming Fang He, Dr. Yiping Wan, Dr. Glen Earthman, and the late Dr. Harbison Pool for their encouragement and professional support.

I

INTRODUCTION

PLANNING FOR EDUCATIONAL SPACE

SCENARIO

The Brown County School District is a steadily growing district located in a suburb of a large southern city. With a current enrollment of 25,000 students, the district is projected to have a continuing annual growth of approximately 2 percent for the next 10 years. The Brown County School District maintains 16 elementary schools, 5 middle schools, and 3 high schools. Half of the 24 school buildings are more than 30 years old and require substantial effort in maintenance. Two new elementary schools were opened last year to replace the two old schools that the school district is keeping for general storage. Capital outlay projects in the next five years include the construction of three elementary schools and one middle school due to enrollment growth. Because of insufficient local resources, the school district is running behind the original building schedule. In addition, the recent education reform bill passed by the state legislature calls for substantial class size reduction at all grade levels. The effect of this class-size mandate puts pressure on the Brown County School District to build more classrooms in addition to its original commitment. With limited fiscal support from the state, the Brown County School District is pushed into a dilemma. While the school district needs to comply with state and federal mandates, it also has to come up with local money to construct one new elementary school and add classrooms at two middle schools. Additionally, because of budget constraints, the school maintenance budget has been reduced tremendously in the past three years, resulting in limited school maintenance activities. Conditions of existing school buildings are deteriorating rapidly.

NEED FOR SCHOOL FACILITIES

School facilities serve as more than just shelters for educational activities (Castaldi, 1994). A well-designed school facility takes into consideration the demands of the educational programs and the needs of its occupants (Chan, 1996a). It enhances school operation and improves attitudes, behavior, and performance of students housed in the facility (Alvord, 1971; Atwater, Gardner, & Wiggins, 1995; Cash, 1993; Chan, 1979, 1982; Cramer, 1976; Earthman, Cash, & Van Berkum, 1996; Hebert, 1998; Hines, 1996; McGuffey, 1972). Studies have also indicated better staff morale and performance in higher-quality facilities (Bowers & Burkett, 1989; Corcoran, Walker, & White, 1998; Proshansky, Ittelson, & Rivlin, 1970). The need for school facilities is further justified by many professional standards for educational administrators. For example, Interstate School Leaders Licensure Consortium standards require school administrators to ensure management of organization, operations, and resources for a safe, efficient, and effective learning environment (Council of Chief State School Officers, 2000). The National Council for the Accreditation of Teacher Education standards highlight the importance of facilities by requiring that school administrators acquire and manage financial and material assets and capital goods and services, allocating resources according to district or school priorities (e.g., property, plant, equipment, etc.) (National Policy Board for Educational Administration for the Educational Leadership Constituent Council, 1995).

SCHOOL FACILITY CHALLENGES IN THE 21ST CENTURY

A U.S. Department of Education report points out that school facility challenges in this century are multifold (U.S. Department of Education, 2000,

April). First, student population continues to rise. By the year 2008, public school enrollment is expected to increase to a record-breaking 54.3 million. Such an increase will have enormous effect on the demand for additional classrooms. Second, small classes and small schools are gaining public support. What this means is that more schools will be built when smaller classes and schools become a legislative mandate. Third, the problem of school-building deterioration due to lack of maintenance is serious. A recent National Education Association (NEA) estimate indicates that $200 billion is needed to build, repair, and modernize public schools (NEA, 2002). Fourth, a school facility challenge of the century is to build schools that can address a broad range of educational needs and that can also serve as centers of community activity (Ryland, 2003). Indeed, the most successful schools of the future will integrate learning communities that accommodate the needs of all community stakeholders.

This book brings into focus the concept of maintenance and renovation as key elements to address the school facility challenges of the 21st century (Chan, 2002). School facility maintenance and renovation have been mentioned in many sources before, but never was their importance stressed enough to demand the attention of educational decision makers. Now is a critical time when facility planners are faced with limited financial resources on one hand and huge facility demands on the other. Revisiting school maintenance and renovation ideas will contribute to resolving critical facility problems. In the years to come, school administrators will have to make the best use of existing school buildings by keeping old school buildings from retiring and preventing usable school buildings from deteriorating rapidly.

REDUCED CLASS SIZE: EFFECT ON FACILITIES

The national trend to reduce class size has a significant effect on educational facility planning. The implementation of a class-size reduction policy substantially reduces the capacity of school facilities and interrupts school construction priorities (Williams, 2002). Many school districts have rented or purchased portable classrooms as a temporary measure to address the problem of classroom shortages (McRobbie, 1997; Roundtree, 1997; Tressler, 1997; Williams, 2002). Others have employed class rescheduling and space subdivision as alternative methods of managing the effects of class-size reduction (Cotton & Linik, 2000; Egelson & Harman,

2000). Recent research (Tressler, 1997) has found that implementation of class-size reduction could be performed more successfully if school districts were given sufficient time to react. Some school districts respond efficiently by using updated information from sophisticated databases (Roundtree, 1997), and some employ a participatory approach by involving administrators, teachers, and parents to resolve facility shortage issues (McRobbie, 1997). The fiscal effect of class-size reduction on school facility planning is more burdensome to school districts of medium and low wealth than school districts of high wealth (Williams, 2002). Consequently, because of serious classroom shortages, some optional, though essential, educational programs are either terminated or moved to other less desirable locations.

FISCAL CONSTRAINTS OF PLANNING NEW FACILITIES

The pressing need for educational facilities is identified in three aspects: new construction, renovation, and maintenance. These needs are the result of continuous enrollment growth, expanded educational programs, reduced class size, poor construction quality, and deferred maintenance. In most states, capital outlay programs are in place as state support for local school facility projects. However, local school districts have to accumulate enough matching funds to participate in these programs. In times of a slow economy, when state funding is tightened, seeking construction dollars through an increase in local taxes or a bond referendum is difficult. In some instances, school districts have to comply with certain state mandates, and special grants from the states to help enforce these mandates are usually insufficient.

In light of tight funding constraints for school construction, some school districts have resolved to pursue the following courses of action:

1. Defer new construction projects;
2. Purchase or rent temporary portable classrooms;
3. Revitalize abandoned facilities;
4. Renovate retiring facilities for continued use; and
5. Refocus on maintenance priorities.

It is anticipated that while new school construction continues in the next decade, much effort will be focused on making the best use of existing facilities by initiating extensive renovation and maintenance projects.

EXTENDING THE LIFE EXPECTANCY
OF SCHOOL BUILDINGS

The general life expectancy of a school building is about 50 years (Castaldi, 1994). If a school building has been well maintained and is in good condition, it can still be used for its intended purpose after 50 years without jeopardizing educational programs. School districts could save a tremendous amount of money if the use of school buildings could be extended for a longer period of time. Without doubt, the life of a school building could be prolonged beyond 50 years, depending on the quality of design and construction, a well-planned maintenance program, and a well-developed vandalism prevention program. In fact, all three conditions originate from the same idea of how to plan and manage school buildings that last.

Quality design and construction refer to designing and constructing school buildings with durability and maintenance in mind. Good specifications of materials and workmanship result in school buildings that last longer and take less effort to maintain. Obviously, higher quality of design and construction will lead to higher contract prices. Expensive school buildings designed and constructed with good quality are justified in the long run by increased years of use and cost savings in efficient maintenance.

A well-planned school maintenance program includes such elements as routine maintenance and preventive maintenance. A reasonable budget needs to be established for the day-to-day operation of school buildings. At the same time, a preventive maintenance program has to be established to schedule anticipated replacement of equipment and building components, such as boilers, roofing system, compressors, and carpet.

A well-developed vandalism prevention program works to deter acts of vandalism by increased supervision. Another aspect of the program is promoting a sense of pride and belonging that students feel about their school. Decreased vandalism means reduced maintenance costs and prolonged life expectancy of school buildings.

THE NATURE OF SCHOOL MAINTENANCE
AND RENOVATION

Many school administrators are not familiar with their roles and responsibilities in school facilities. Their lack of knowledge and skills in this specific area puts them at a disadvantage when they are faced with the school board and the public for answers about school maintenance and renovation. Because of limited financial resources, abandoned or retiring facilities will be seriously considered for renovation and reuse. Resources for new facility construction projects may be redirected to maintenance and renovation use so that tax dollars can be stretched to prolong the life of old school buildings. At the same time, attention will be given to the care of existing school buildings that are still in good condition so that they can last longer. What this means is that school administrators need to realign their school facility plans to include more renovation projects as priorities. School administrators also need to gather valid data to support a reasonable school maintenance budget. School board members and the public need to understand that an increased budget for school maintenance is a long-term investment. Current and potential school administrators need to be well prepared to face the facilities challenges of the 21st century.

This book addresses the school facility maintenance and renovation issues that school administrators are confronting right now. A macro approach will be taken to explore the facility planning effort at the district level, and a micro approach will be employed to discuss the specific roles and responsibilities of school building administrators in managing facilities. The book is based on real experiences with specific examples. Useful tools are incorporated to facilitate administrators in performing their daily duties. This book is not only essential for instructional use as a textbook for leadership preparation programs but as a basis for facility in-service workshops. In addition, this book will help current school administrators, school board members, and the general public understand the importance of school maintenance and renovation and how they can contribute to making the best use of facility resources.

THE VALUE OF SCHOOL MAINTENANCE AND RENOVATION

The purpose of school maintenance is to prolong the usable life of school buildings by keeping them in good condition, while renovation gives school buildings a new life through face-lifting, safety checking, and educational function accommodation. Both maintenance and renovation programs are developed to achieve the same goal of saving educational resources for the school district by making the best use of available school buildings.

Maintenance

A well-developed maintenance program helps slow down the school building deterioration process. It provides a healthy environment for teaching and learning. A school building neatly maintained inside and out sends a positive image to the public that the school district cares.

Renovation

Renovation is seen by the community as an effort by the school district to uphold community culture and heritage. It saves tax dollars by preserving the working sub-systems of an old building, while replacing its substantially damaged structure for safety. Additionally, renovation usually results in redistribution of spaces to accommodate new programs. Even though some simple renovation projects are basic cosmetic dress-ups, they serve the purpose of winning public support.

SUMMARY

This chapter identifies the needs for educational facilities and outlines a picture of the school facility challenges of the 21st century. The deteriorating conditions of many school facilities, together with limited fiscal resources, lead to educational administrators' interest in school maintenance and renovation programs. The nature and value of school maintenance and renovation programs are also explored.

EXERCISES

1. Identify the school facility needs of your school district in the next five years by category of project: new construction, renovation, and maintenance.
2. How does class-size reduction affect school facility planning in your school district? Assess the extent of the effect by interviewing your superintendent.

3. How do school maintenance and renovation programs help the master planning for facilities in the Brown County School District in the scenario?
4. Compare the average per-square-foot cost of new construction and renovation of schools in your geographical area.
5. As a school administrator, develop a plan to fight school vandalism. You may want to establish a committee to involve all the stakeholders.
6. Prepare a presentation to the school board in support of school renovation and maintenance programs. Provide data to support your position.

WHAT RESEARCH HAS TO SAY ABOUT SCHOOL MAINTENANCE AND RENOVATION

SCENARIO

The Davis School District is located in a suburb of a mid-sized southern city. With current student enrollment of 50,000, it is experiencing a steady growth of 1 percent. The school district, working closely with the state department of education, has identified three new elementary schools to be constructed in the next five years. However, recent class-size reduction legislation about puts the district's facility plan off track. The school superintendent has asked the School Facility Planning Office to conduct a thorough evaluation of what class-size reduction really means to the facility priorities of the Davis School District. Specifically, the superintendent is willing to explore the option of converting the construction dollars for one of the new elementary schools to the renovation of three retiring elementary schools. The superintendent's idea may make sense because the three retiring elementary schools have a capacity of 1,000 students each. If substantial dollars are approved for renovation, these schools will be completely modernized to meet the needs of current educational programs. This will reduce tremendously the pressure on facility demand as a result of class-size reduction implementation initiatives. At the same time, the superintendent is planning to increase substantially the school maintenance budget to keep all the school buildings of the Davis School District in good condition. The superintendent is preparing to bring his renovation and maintenance project proposal to the school board for approval. He is looking for good justifications for his proposal through a literature review.

INTERFACE OF THE LEARNER AND THE LEARNING ENVIRONMENT

Cunningham (2002) discusses interaction between the student and the learning environment. He claims that environment influences student behavior in school. In reviewing environmental documents, Schneider (2002) found that clean air, good light, and a quiet, comfortable, and safe environment contributed to student learning. A review of the literature shows that many studies have been conducted examining student achievement in old versus new school buildings. Most of them found significant differences in student achievement between old and new school buildings. However, very few studies have been performed on the effect of school renovation and maintenance on student performance. Chan (1980) compared the achievement of seventh-grade students in new, old, and renovated school buildings in Georgia. Students in new school buildings had higher achievement than those in old and renovated school buildings and those in old school buildings did not perform as well as those in renovated buildings. In a school revitalization demonstration project in Washington, D.C., Berry (2002) witnessed how an improved school environment contributed to higher levels of educational performance. Maxwell (1999) studied school renovation projects in one district and concluded that a positive relationship existed between upgraded school facilities and math achievement.

SCHOOL RENOVATION CONSIDERATIONS

Organization of planning procedures starts when the decision is made to renovate a school building. Peters and Smith (1998) developed a step-by-step plan for school renovation projects:

1. Evaluate your existing resources;
2. Establish the district's educational philosophy;
3. Compile key facility program information;
4. Recognize the effect on existing facilities and stakeholders during construction; and
5. Have reasonable expectations.

Earthman (1994) and Hill (2000) proposed similar school renovation procedures. Earthman also recommended a renovation evaluation process as an essential component of the renovation plan. Henry (2000) and MacKenzie and Phillips (1991) identified needs assessment as the most important phase of a school renovation project. The latter developed a survey instrument to evaluate the condition and maintenance needs of renovated school buildings. The concerns of planning a school renovation project were identified by Coffey (1992) as: structural soundness, program support, site, and cost. He also listed specific renovation elements to include the development of educational specifications, attention to site condition, consideration of playground areas, importance of the exterior appearance of the school buildings, space utilization, condition of mechanical and electrical systems, importance of energy efficiency, development of barrier-free environments, treatment of thermal environments, consideration of acoustics, management of visual environments, selection of furniture and equipment, and attention to aesthetics.

SCHOOL PRINCIPAL AND RENOVATION PROJECTS

Whether the principal is physically present at a school renovation project or not, he or she is involved in the planning and construction process. This is based on the facility planning principle that occupants are most knowledgeable about what suits them best. The benefits of involving principals in school renovation projects are confirmed by Brent and Cianca (2001), Cianca and Brent (2002), and Futral (1993). As expressed by Futral, involving school principals in renovation projects allows them the opportunity to influence quality and usefulness of renovated space, as well as the values expressed by school environment. However, Cianca and Brent (2002) expressed concern that principals excited about renovation in their schools could easily step over their legitimate role as principals. They also questioned principals' background and training in managing school renovation projects. Chan (1999) urges that principals identify their roles and follow the line of authority in the management of school construction projects. Brent and Cianca (2001) and Earthman (1998) discuss the nature and extent of principals' involvement in school renovation projects. Earthman stresses that the pitfalls of school renovation projects need to be explained to principals to enable them to make more informed decisions in the interest of school safety and educational programs.

SCHOOL MAINTENANCE AND HEALTHFULNESS

School buildings have to be maintained satisfactorily on a daily basis to meet state and local health standards. This includes basic cleaning, controlling for air quality, providing clean drinking water, keeping sanitary conditions for food preparation, and managing pest control. As health standards get more stringent, the demand for compliance is greater on maintenance staff. Shaw (2000) points to the Chicago public schools as an example of a system that successfully met high health standards. As Shideler (2001) addressed it, "A clean school is a healthy school." The main purpose of the school maintenance program is to provide a clean and healthy environment to support teaching and learning as evidenced by less absenteeism, illness, and injury. Synthesizing studies pertaining to facilities, student achievement, and student behavior, Lemasters (1997) identified maintenance as an independent variable affecting student achievement and behavior. In discussing indoor air quality control, Dunklee and Siberman (1991) put it in plain language, "Healthy buildings keep employees out of bed and employers out of court." They recommend specific procedures to ensure healthy indoor air quality. The Responsible Industry for a Sound Environment (1999) particularly stresses the importance of pest control to maintain a healthy school environment. It claims that pests do more than disrupt the learning environment; they pose serious health threats to children. Jacobs (1995) and Healthy Schools Network Inc. (2000) drew the attention of administrators to maintain healthy school conditions during renovation to protect the staff and students. Finally, the key to maintaining high standards of healthy conditions in schools has been identified as extensive staff training and a sizeable investment in building repair (Kennedy, 2000; Shaw, 2000; Shideler, 2001).

IMPROVED SCHOOL ENVIRONMENT
AND STAFF AND STUDENT MORALE

The physical conditions of an educational environment affect the morale of teachers and students. Students in better-equipped environments have more positive attitudes toward their schools (Chan, 1982). To study teacher and student morale and its connection with school physical environment, Tanner and Morris (2002) proposed a practical model of a survey questionnaire and statistical approach. They anticipated that improved morale could be achieved when school appearance and school design were improved. Dawson and Parker (1998) conducted a study to examine the effects of facility renovation on faculty morale. The findings indicated that faculty expressed their frustration with the change but that overall morale appeared to be high as a result of the renovations. An interesting study was conducted by Stapleton (2001) investigating the perceptions of parents, staff, and students toward school climate between old and new school buildings. The study showed that the perceptions of parents, staff, and students toward school climate in the new building were not as high as they were in the old building. Acclimation to a new school building was the explanation.

SCHOOL MAINTENANCE AND FACILITY LIFE EXPECTANCY

School buildings carefully planned and built can normally last for 50 to 60 years (Castaldi, 1994). However, school buildings deteriorate every day because of wear and tear and the functions of mother nature. "As building materials age, their structure and systems weaken and deteriorate" (Vasfaret, 2002, p. 6). School buildings with ineffective maintenance run down quickly to complete failure. Only proper and timely maintenance helps to slow the building deteriorating process and ensure the continued use of a facility (Rabenaldt & Velz, 1999). As a result of good maintenance, the usable life of a school building is extended and the capital outlay investment value is stretched. In analyzing the procedures for opening new school buildings, Chan (2000) stresses that principals must take the lead in developing effective maintenance programs that will prolong the life of the school buildings. Simko (1987) outlines the stages of an effective maintenance program and concludes that maintenance offered the benefits of longevity and reliability of physical plant components.

SCHOOL PREVENTIVE MAINTENANCE

School preventive maintenance takes a proactive approach by routinely checking and addressing the potential problems of school facilities. The life expectancy of school equipment and operating systems can be projected by past
experience. These equipment and operating systems need to be replaced or reworked before their time runs out. Vasfaret (2002) described the concept of
preventive maintenance as fighting time and elements. In fact, the ability to control deterioration makes the job of school administrators less distressful. Since
maintenance is important (Moore, 2002) and unavoidable (Pittillo, 1993), it
makes good sense to develop a plan to identify potential facility problems and
work on them before they become real problems. Emergency repairs through
reactive maintenance (after the breakdown) result in high costs, while proactive
maintenance (before the breakdown) saves money by cutting down on crisis
spending (Simko, 1987). Sharp (1992) expands the concept of preventive
maintenance to focus on health and safety issues in school. He recommends
that school administrators frequently walk through the school building with the
purpose of identifying potential health and safety risks. Maintenance work of a
preventive nature frequently performed by school districts includes reroofing
(Jozwiak, 1998; Kalinger, 1998), plumbing retrofitting (Westerkamp, 2000),
and landscape improvement projects (Spitz, 2001).

IMPROVEMENT COST AND LIFE-CYCLE SAVING

When an investment is made to improve a school building, a life-cycle cost
analysis is often made to better understand the efficiency of the anticipated outcome. A life-cycle cost analysis ties the initial costs to the operating costs of the
proposed improvement to yield a better picture of the real incurred expenses
over a period of time. The analysis assists administrators in deciding whether
the proposed investment is worthwhile in terms of the time it takes to payback.
Consideration of the life-cycle cost factor is beneficial to the planning and maintenance of school facilities (Hoffman, 2002). To examine the initial costs versus
operating costs of improvement projects, Chan (1982) studied the payback
years of an initial investment in 14 school improvement projects in South Carolina. Results indicate that only 10 schools were able to payback the initial investment in the anticipated years. A similar study was conducted by Bennett

(1983) to investigate the paybacks of energy conservation initiatives in Michigan schools. The findings showed substantial savings for the project schools over 20 years. School improvement projects commonly assessed by life-cycle costs include modern lighting (Illuminating Engineering Society of North America, 2000), energy conservation (Fuller & Petersen, 1996), floor finishes (Moussatche, Languell-Urquhart, & Woodson, 2000), and roofing system (Wright, 1996). Gardner (1977) notes that a cheaper model costs more per hour of use than a more expensive, but longer lasting, model. Holt & Kirby (1997) concur that a low bid too often equates to low quality, high maintenance, poor performance, and a short life. They urge that school districts prepare bid specifications with life-cycle costs in mind. In addition, handbooks and guidebooks have been written on the step-by-step process of conducting a life-cycle cost analysis (Kirk & Dell'Isola, 1995; Mearig, Coffee, & Morgan, 1999).

SUMMARY

The literature reviewed on school maintenance and renovation in this chapter represents highlights of the existing major work on school maintenance and renovation. Some studies take a practical approach by addressing the "know how" of performing the tasks. Many are research oriented and are supported by data collected in the field. Most of the literature reviewed shares similar points of view. They reinforce one another to help clarify controversial issues in school facilities planning. The topics reviewed in this chapter are summarized in the following points:

> *Interface of the Learner and the Learning Environment*: Most of the studies reviewed indicated significant improvement in student achievement in environments of better quality. However, research on school maintenance and renovation is scarce. More experimental studies with stringent controls are needed.
>
> *School Renovation Considerations*: The literature reviewed in this section is focused on the step-by-step procedures of managing a school renovation project. Needs assessment and evaluation components of the renovation process were highlighted. The general impression is that school renovation projects must be well planned before they are started, especially those ongoing during school hours.

School Principals and Renovation Projects: Current literature confirms that school principals are essential stakeholders in the planning process of school renovation. They need to be well informed and prepared when it comes to decision making. School principals need to be aware of their roles and responsibilities in renovation projects.

School Maintenance and Healthfulness: A school maintenance program is aimed at providing a healthy environment for teaching and learning. The literature provides guidelines to assist in maintaining a healthy school. It is clear that health standards have no room for compromise.

Improved Environment and Staff and Student Morale: Studies in this area are scarce and contradictory; more research is needed. Particular attention has to be paid to the control of independent variables that could affect morale.

School Maintenance and Facility Life Expectancy: The literature cited in this section is not research-based. However, all the studies point to the understanding that investment in school maintenance will prolong the usable life of a school building.

School Preventive Maintenance: The literature cited in this section relates to the professional experiences of the authors. The message is clear that school maintenance is unavoidable and that preventive maintenance saves money in the long run.

Improvement Cost and Life-cycle Savings: The facility improvement projects cited in this section were designed with the anticipation of saving money after the payback years. Most of the investments were paid back as designed. Administrators must plan school retrofit projects carefully.

The literature reviewed in this chapter reveals that school maintenance and renovation are performed not only to save money in the long run but also because of a continuous commitment of providing a positive environment for teaching and learning. Investment in school maintenance and renovation is fully justified in the literature. The long-term benefits cannot be underestimated.

EXERCISES

1. As director of school maintenance, prepare a 10-minute speech for the school board meeting to protect the threatened maintenance budget. Draw evidence from recent literature to support your argument.

2. Should principals be involved in the management of school renovation projects? Discuss in your opinion the extent of involvement.

3. Propose an operational plan to efficiently manage your school building. Highlight the benefits of life-cycle savings and building durability through your review of the literature.

4. Some literature has related student achievement to school renovation and maintenance. Do you see the connection?

5. Is preventive maintenance costly? Justify your preventive maintenance plan by citing from recent literature you reviewed.

6. How do you maintain the healthy condition of a school building? What have you learned from the review of literature?

7. As project administrator of a school renovation project, how do you maintain a safe environment for the faculty and students who are still using the school building? Support your ideas with the current literature.

8. Check with your local health department for a copy of the health standards your school has to meet. Check all items in your school by using the standards. Develop a list of compliance items and a list of noncompliance items.

SCHOOL
MAINTENANCE
AND RENOVATION:
CONCEPTS AND
MODELS

SCENARIO

The Brown County School District has a newly elected school board consisting of seven members, two of whom are encumbered and five of whom are new. Most of the school board members do not have any training in school planning and construction. To balance the budget in this tight economic year, most of them, particularly the new members, are leaning toward cutting school maintenance budgets enormously and postponing major school renovation projects. As the school superintendent recalls, the former school board cut maintenance and renovation budgets two years ago. Many school maintenance projects and renovation proposals were delayed as a result of that decision. Many schools are now in desperate need for immediate repair. Therefore, the superintendent does not want to trim the maintenance and renovation budget any more this year. To change the minds of the new school board members, the superintendent has asked the director of school facilities to prepare a three-hour workshop to update the new school board on the current conditions of school buildings in the Brown County School District and to brainstorm with the board members about the basic concepts and possible models of school maintenance and renovation. The workshop will conclude with concrete recommendations from the director of school facilities about how effective school maintenance and renovation programs could meet the facilities needs of the school district.

WHAT DOES SCHOOL MAINTENANCE AND RENOVATION REALLY MEAN?

School maintenance is a program to maintain existing school facilities to create a positive environment for teaching and learning. To achieve this purpose, an effective school maintenance program has to achieve four goals:

1. To maintain the school in such clean and tidy conditions as meet health standards;
2. To maintain the school in excellent operational conditions to meet the physical needs for teaching and learning;
3. To maintain the school in a timely manner to prolong the usable life of the systems; and
4. To maintain the school efficiently by making use of all available resources.

A school maintenance program needs to be initiated at the district level, setting goals and providing guidelines for individual schools to achieve their goals. Each school needs to compose its own maintenance plan to achieve the district goals.

A school renovation program aims at providing a school building with a new life by refreshing its appearance or improving its operational functions. An effective renovation program has one or more of the following goals in mind:

1. To make the appearance of the school building aesthetically more attractive;
2. To improve the energy efficiency of the school building;
3. To replace the worn-out building operation systems; or
4. To accommodate the new educational mission of the school building.

Renovation plans for school buildings in a school district are prioritized in the central office according to the physical needs of the schools and the directions for change in educational use. Renovation projects are more than cosmetic touch-ups. They serve to improve the efficiency and life expectancy of school buildings.

WHAT ARE WE ACHIEVING IN SCHOOL MAINTENANCE AND RENOVATION?

When a school district keeps its school buildings in good condition by developing effective maintenance programs and well-planned renovation projects, it sends a strong message to the community that it cares about education. This message is unbelievably powerful in that it never fails to gain community support. The community is proud of well-maintained schools in its neighborhood. Parents are pleased to see that their children go to a well-maintained school. When a renovation project goes on in a school, community members cannot wait to see the appearance of the completely renovated school. In fact, budgeting money for proper school maintenance and renovation help develop good community relations because the community likes to see tax dollars spent in the improvement of the schools their children attend. To many elderly community members, school maintenance and renovation are a means to preserve the cultural heritage of their community. In fact, this is the best way to say, "We spent your money wisely." Well-maintained schools and well-planned renovation projects speak for themselves by presenting an attractive appearance and showing clear evidence of prolonging the lives of school buildings. As a result, tax dollars can be saved by having to construct fewer new school buildings. Effective school maintenance programs and well-planned renovation projects save millions of dollars of construction money for instructional use.

MAINTENANCE CONCEPTS

The following school maintenance concepts have varying degrees of value; however, they provide an opportunity for administrators to get out of the traditional paradigm to explore new alternatives.

Run-to-Failure

The run-to-failure concept of maintenance is not a recommended maintenance approach because it is passive in nature. It is mostly adopted by school districts that have a minimum maintenance budget. The maintenance departments are usually so understaffed that all they can handle are requests from schools for emergency repairs; preventive maintenance work is not affordable or planned. Schools operate on this kind of tight maintenance budget at the expense of the future.

Maintenance Free

Maintenance free usually means it requires no maintenance. Designers should have maintenance in mind when deciding on educational features in a school building. Items with no or low maintenance save tremendously in long-term operating expenses. Sometimes maintenance-free features do incur additional costs up front, but the life-long savings justify the investment. For some building designs, maintenance free does not carry the life-long guarantee. It may simply mean that buildings do not require frequent maintenance.

Vandal Proof

The term *vandal proof*, strictly speaking, describes a condition that does not exist. The fact is that no matter how much thought has been put into the design of a building to prevent vandalism, the building is still vulnerable to vandalism. Therefore, the term cannot be taken in its literal sense. What vandal proof really means is that certain design features in a school building cannot be easily vandalized and are designed to fight vandalism. The term *vandal proof* is often associated with durability. As long as a building design can stand a certain degree of deliberate attempts of destruction, it is acceptable as a vandal proof design.

Safe Buildings

In designing and maintaining school buildings, school administrators have the primary responsibility of providing a safe environment for teaching and learning. Safety does not only refer to building structure; it also includes high-quality indoor air, lead-free drinking water, fire safety features, emergency utility cutoffs, and well-maintained indoor and outdoor equipment. Safe buildings are periodically put to the test with fire and tornado drills. Safe school buildings are designed and maintained according to building codes, fire codes, and public health regulations.

Routine Maintenance

School buildings need to be maintained with a routine schedule so that specific jobs can be assigned to responsible staff with specific time frames. Routine maintenance items, such as trash removal, vacuuming, waxing, minor repairs, HVAC system management, bathroom sanitation, and lawn mowing, are needed to achieve a high quality of tidiness and cleanliness in the school building. The central office maintenance department will work with school principals on staffing and budgeting issues. Some school districts have experimented with contract service with professional cleaning companies to perform routine school maintenance after school hours.

Maintenance Liability

When school building systems are not efficiently or effectively designed, it takes an enormous amount of time and effort to keep them running. As the systems age, efficiency depreciates even faster and the systems become a maintenance liability. Some design features are maintenance liabilities because of their high operational expense. For example, some school districts prefer to replace hard-to-operate heating and air-conditioning systems over continuing to pay for high energy consumption.

RENOVATION CONCEPTS

When the decision is made to renovate a school building, many renovation concepts will come to the drawing board. Renovation can be as simple as addressing cosmetic issues or as complicated as reworking the available space to serve a better purpose. The shortcomings of renovation are the hidden uncertainties that could emerge and the limitations that the existing structure creates. But all school renovation projects offer exciting challenges to facility planners. The following are some renovation concepts to be considered.

School Aesthetics

School aesthetics are a basic element of a school renovation project. A school has to be more aesthetically pleasing after renovation. This is what the community expects. Renovation dollars need to be spent where people can observe improvements. School improvement items, such as repainting, new floor coverings, new ceilings, and new lighting, will certainly give the interior spaces a fresh, new look. Exterior aesthetic work can include new brick veneer, new windows, repavement of parking spaces and driveways, landscape redesign, new school signage, and new exterior lighting.

Space Utilization

One aspect of school renovation is to examine the space utilization of the school in renovation. The existing school is obsolete not only because of limited square footage but also because of inadequate utilization of the square footage. Therefore, a redistribution of space is necessary to accommodate the needs of the new educational program. This often involves the removal and erection of walls, reassignment of space functions, and conversion of existing spaces for other uses. The purpose is to realign the existing square footage of the building to serve a better educational purpose.

Hidden Conditions

One of the biggest problems in school renovation is to deal with hidden conditions. Many old school buildings have been reworked many times before. It is very difficult to track down specifically what has been done since the school was first opened. Even as-built drawings do not necessarily display all the changes as a result of change orders. Therefore, experienced planners tend to reserve more contingency dollars in school renovation projects to manage problems associated with hidden conditions in old school buildings.

Structural Limitations

School renovation projects are very much limited by the structure of the existing building, particularly the load-bearing walls and the supporting columns. It is easier to convert an old gymnasium into a media center than to develop four existing classrooms into a construction shop. Renovation projects are also limited by the configuration and the expansion capacity of the existing facilities. Some changes cannot be made without incurring huge expense.

Functional Renovation

School renovation work in this area addresses building problems relating to the malfunctioning of the major systems, such as structural, mechanical, electrical, and plumbing systems. These systems are essential to the daily operation of a school. Money spent in functional renovation will help improve the functional operation of the school building. School buildings will last longer because of extensive work done in functional renovation.

MAINTENANCE MODELS

Traditional school maintenance usually takes the fix-it-as-it-breaks approach. But school administrators find it more economical to invest in the beginning by initiating preventive maintenance programs that address failing building systems before they come to complete failure. Then came the idea of contracting outside professional maintenance companies to perform school maintenance work. On the other end of the extreme, some large school districts keep their own in-house construction crew to work on all school maintenance and minor renovation projects. The different school maintenance models are further explained.

Corrective Maintenance

A corrective approach to school maintenance covers the basics of a school maintenance program. It addresses all the repair needs requested by the schools. This corrective approach does not call for any checking of services before system failure. The approach, though inexpensive, poses concerns about the shortsightedness of investment in facility management.

Preventive Maintenance

School preventive maintenance takes a proactive approach of addressing antic-ipated maintenance problems before the real problems occur. There are several aspects of this approach. First, it projects the life expectancy of building com-ponents and replaces the going-to-be-worn-out parts before they completely run down. Second, a site-based facility inspection system is developed to detect faulty or soon-to-be-faulty facilities and equipment (Earthman, 1994). Third, a routine schedule is developed to service the essential parts of school equipment to ensure efficiency and durability. Preventive maintenance is the preferred approach to school maintenance. But, unfortunately, preventive maintenance is always one of the first budget items to go in fiscally tight years.

Contract Maintenance

School maintenance is traditionally performed by the school district maintenance department that keeps its own maintenance staff. School custodial staff are employed by school principals based on the recommended list supplied by the maintenance department. Contract maintenance is a new model of maintenance being tried by some small to medium-size school districts. When a school district contracts with an outside company to perform maintenance work, the school district does not employ its own maintenance staff. The contracted company is responsible for all the routine cleaning work and any needed repair work as requested on call. Some school districts find it convenient, while others are ready to go back to having their own maintenance staff (Kowalski, 2002).

RENOVATION MODELS

School renovation is a general term covering all the construction work done to an existing school building to improve the facilities for teaching and learning. The extent of a renovation project is dependent upon the size of the assigned budget and the identified priority of needs. Therefore, a renovation project could be surface-oriented or function-oriented, or both. Different renovation models, such as rehabilitation, remodeling, and modernization, are explained in the following section in more detail.

Rehabilitation

School rehabilitation as an aspect of school renovation refers to improvement of existing facilities so that an old school building can be restored to its original shape by face-lifting and the replacement of worn-out systems. The purpose of rehabilitation is to recover the outlook and to reinstall the proper functioning of an existing old building (Castaldi, 1994). In terms of the scope of renovation work, rehabilitation involves the least amount of construction work.

Remodeling

Remodeling as a component of school renovation carries a face-lifting aspect in general sense. It particularly features work in moving wall partitions to create new spaces to suit new functions. As a result of remodeling, the square footage of a school building may not change, but the available indoor spaces may be relocated and redistributed in a way that will meet the needs of the current occupants. New rooms may need to be renumbered and locks rekeyed.

Modernization

Compared with rehabilitation and remodeling, the scope of work is the most extensive in school modernization. It basically covers all the work mentioned in rehabilitation and remodeling. Unlike rehabilitation, a distinct characteristic of modernization is to improve the functioning of the facility so that it will operate more efficiently. Upsizing the HVAC system, replacing inefficient equipment, increasing storage capacity, upgrading safety devices, replacing the boiler, removing asbestos, accommodating the handicapped, and improvements to meet new codes are specific examples of modernization work.

In-House Construction

Some large school districts keep an in-house construction crew for small construction projects. They find that keeping an in-house construction crew saves money for the school district. In addition, it is a convenience because it is very difficult to find contractors to bid on small school projects. Many school districts have their own woodshop, AV repair shop, blind-making shop, auto mechanics shop, and warehouse to store basic construction materials. Large school districts traditionally hire their own architects on staff for the design of small school renovation projects.

SUMMARY

The concepts and models of school maintenance and renovation are explored in this chapter. It is clear that some concepts are more practical than others. The choice of approach to school maintenance indicates the management concept of the decision makers. School renovation can be very involved, but school building safety needs to be the priority issue to be addressed. A common factor to be considered in the final selection of maintenance and renovation models is the size of the school budget. Since maintenance and renovation budgets have been minimal, all effort has to be exerted to protect these budgets from further cuts. School decision makers need to be reminded that school maintenance and renovation mean long-term investment in educational facilities.

EXERCISES

1. What should be included in a school renovation project? What are your justifications?
2. In a tight school budget year, prepare a list of reasons in support of cosmetic items in your school renovation project.
3. Explore the pros and cons of contracting school custodial services by interviewing two directors of school maintenance. Do you think contract custodial services will work in your school?
4. Why is the school maintenance budget always the first to be cut? What can you say to protect the school maintenance budget?
5. In making decisions on school renovation projects, how do you balance between face-lifting work and functional work?
6. Some vandal-resistant design elements in school buildings are affordable. Compose a list of these school design items from a literature review and personal contacts.
7. Identify the facility deficiencies of your school by organizing a school facility committee. Prioritize the list of deficiencies to get ready for upcoming school renovation.
8. Give some examples of school maintenance liabilities and suggest how they can be avoided.

II

SCHOOL
MAINTENANCE

4

PLANNING FOR SCHOOL MAINTENANCE AT THE DISTRICT LEVEL

SCENARIO

The Thomason County School District is a school system in the Midwest with a total pupil population of 50,000 housed in 8 high schools, 12 middle schools, and 28 elementary schools. The district is experiencing a slight growth of approximately 400 pupils per year. With the current growth rate, the district has projected the need for the construction of one new high school, one new middle school, and three new elementary schools in the next 10 years. To Mr. David Hill, the new school superintendent, the construction of new facilities to meet growth is not the pressing issue. What worries him more is the deteriorating condition of the existing school buildings. More than half of the school buildings in the Thomason County School District are between 30 to 40 years old. Attention is drawn to properly maintaining these facilities so that they can be used to serve education purposes for an extended time. Because of budget constraint, school maintenance for years has followed the basic, passive approach of fix-it-as-it-fails. Not much preventive planning was done nor was it affordable. Mr. Hill, with a facility construction and management background, has taken a strong stand in properly maintaining the school buildings and is planning to present a facility maintenance report to the school board for approval. He has asked Mr. Rogers, director of the maintenance department, to completely evaluate the management of his department with a focus on identifying facility maintenance needs, a preventive school maintenance schedule, and the resources necessary to implement a successful school maintenance program. Mr. Hill also suggested to Mr. Rogers that he consider examining the efficiency and effectiveness of his department as well as reorganization initiatives.

PLANNING FOR SCHOOL MAINTENANCE

Planning for an effective school maintenance program at the district level is most important because the purpose of the program is to safeguard the school district's large investment in school buildings. When new school buildings are completed and turned over to the school district, they begin their deterioration process. An effective maintenance program will help slow down the deterioration speed, keep up the school's appearance, and prolong the useful life of the school building. To develop an effective maintenance program, realistic goals must be established. Some examples of school maintenance goals include:

- Maintaining a clean and healthy environment for teaching and learning
- Developing and implementing an effective preventive maintenance program for all school buildings
- Developing and implementing an effective energy conservation program for all school buildings
- Developing and implement an effective proactive maintenance program for all school buildings.

Based on the established goals, specific objectives can be developed to reflect the related activities. To develop school maintenance objectives, criteria must be included for valid measurement. Some examples of school maintenance objectives include:

- Maintaining a clean and pleasant school environment at all times
- Keeping a healthy school environment to meet state health standards

- Conducting a reliable energy audit for conservation effectiveness
- Completing a school-by-school facility inventory for planning use
- Evaluating the effectiveness of the recently completed boiler replacement projects.

To meet these school maintenance objectives, maintenance activities need to be developed, and a budget has to be proposed to include staff and material resources to support these activities. For the maintenance program to be successfully implemented, specific step-by-step procedures need to be outlined. Supervisory activities should be included as part of the implementation plan. An evaluation component is essential to measure program effectiveness as part of the school maintenance program.

BUDGETING FOR MAINTENANCE RESOURCES

The budget for school maintenance consists of two parts: individual school budgets and district-level budgets. Recent statistics show that the national average of school maintenance expenditures is 7.4 percent of the total school district expenditure (National Clearinghouse for Educational Facilities, 2003).

The school maintenance budget consists of a section for custodial staff and a section for equipment and supplies. The budget for both sections is calculated based on pupil population or school building square footage (Thompson & Wood, 2001). Since the formulae are fixed at either the state or district level, calculations for the budget for each school should be simple.

To develop a school district-level maintenance budget, the following items should be considered: staffing, maintenance equipment and supplies, office equipment and supplies, overhead expenses, vehicle expenses, staff development, and projected preventive maintenance projects. The last year's budget could serve as good reference to compose the new budget. But budget decisions have to be based on data-driven projected needs of the school district in the coming year. Each line of the maintenance budget request needs to be carefully documented. It is a pity, however, that maintenance budget requests usually suffer the most severe cuts even in growing school districts.

THE SCHOOL DISTRICT MAINTENANCE OPERATION

School district maintenance is an enormous operation. The larger the school district, the more complicated the maintenance task. Small school districts have fewer schools but still have to deal with the same maintenance problems. A typical school maintenance department at the district level is made up of divisions such as electrical, plumbing, carpentry, HVAC, janitorial, grounds, flooring, glass, keys and locks, roofing, painting, and AV repair. Auto maintenance is sometimes combined with the transportation department, which has the responsibility of school bus maintenance. Each division is led by a foreman. The number of staff in each division depends on the number of maintenance requests the division receives. All foremen report to the director or assistant director of maintenance. Some school district maintenance departments even operate their own woodshop, welding shop, shutter center, and construction crew. The rationale is to keep the small manufacturing and construction projects at the lowest cost by employing school district staff to work on the projects. The office of the maintenance department (see organizational chart in appendix A) usually maintains staff consisting of the director and one or more assistant directors, one or more administrative assistants, and one or more supervisors. While the director and assistant director(s) have the overall planning responsibility for school maintenance, the administrative staff keeps the office running through coordination and record management. The supervisors review maintenance requests every day and assign work priorities to the foremen of the divisions. A warehouse is usually kept by the maintenance department to store materials and parts needed for school maintenance. The maintenance warehouse is headed by a warehouse supervisor who is responsible for storage and inventory of items needed for maintenance use. For the maintenance department to function, a fleet of vehicles has to be maintained to deliver services to the many schools and locations of the school district. The type and number of vehicles needed are determined by the nature of the work the vehicles are used for. In addition, tools and special equipment are also needed for use in school maintenance.

When maintenance requests are received, those addressing safety concerns, particularly those marked urgent, are given priority. All other requests are handled on a first-come-first-serve basis. The time it takes for the maintenance department to respond to requests depends on the size of the maintenance staff and the complexity of the request. Small school districts do not have the luxury of divisional organization. The few maintenance staff members they have are proficient in many maintenance trades. Because of the size limitation, small school districts have a greater tendency to contract a portion of their school maintenance work.

USE OF TECHNOLOGY IN SCHOOL MAINTENANCE

Computer technology has helped maintenance departments in many aspects of their daily operation. First, the construction data of all the school buildings can be stored in the computer and retrieved in any format as needed. The history of major school renovations and maintenance can be kept in the same files. All new construction drawings can be stored on discs and old drawings can be scanned for easy storage. The entire archive can be computerized. Second, programs are available for electronic submission of maintenance requests from schools to the maintenance department. Maintenance requests can be prioritized and tentative dates set up for repair work. An electronic response can be sent back to the schools for their records. Third, work schedules and work hours of individual maintenance staff can be recorded in the computer for future reports. Any personnel data can be electronically filed. Fourth, computer technology is effectively used for maintenance warehouse inventory because all checked-in and checked-out items can be accurately recorded.

In addition to the maintenance department, computer technology is used extensively in the operation of school buildings. In many schools, computerization is used in programming for heating and air-conditioning systems, boiler and plumbing systems, intrusion alarm systems, fire protection systems, surveillance systems, and all internal and external networking systems. School maintenance staff can no longer maintain a school building without knowledge of the technology setting in the schools. School districts either have to retrain their maintenance staff for technology knowledge in their fields or contract the technology portion of maintenance work to firms that can do the work.

SCHOOL MAINTENANCE PRIORITIES

When school maintenance requests are received by the maintenance department, they need to be prioritized so they can be handled in a timely manner.

In making decisions of prioritization, the following principles can be followed:

1. Safety. All maintenance requests relating to safety issues have to be treated as an emergency and have to be addressed immediately, even if maintenance staff have to be pulled from other assignments.
2. Deadline. Some maintenance requests have a specified deadline. The work has to be done by the requested time for certain events. Missing the deadline could incur serious consequences.
3. Length of time. How much time it will take to complete the requested work is an important consideration in determining priorities. It makes sense to schedule some of the nonurgent and time-consuming work for the summer so that more time can be reserved to take care of more urgent requests during the school year.
4. First-come-first-serve. First-come-first-serve, like take-a-number, is a basic principle that no one can dispute. Prioritization can be done according to the time and date maintenance requests are received.
5. Nature of requested work. Some requested work has no urgency in terms of safety or meeting deadlines. But some of these problems, if not addressed in a timely manner, could be intensified and cause other related damages. Then, the repair work could be costly and difficult.
6. Geographic location. Geographic location is an important factor in considering work priorities. Even though a request may have come in later than other requests, it makes sense for the maintenance staff to take care of it while they are in the same area doing other work. Considering geographic location in scheduling is a way to make the best use of maintenance resources.

When school principals submit maintenance requests, they are anxious to know when the repair work will be done. Therefore, responding to principals' requests by returning tentative work schedules is a responsible way to handle business. Principals need to be notified of any changes that could delay the original schedule. To understand the nature of urgent requests, the maintenance department needs to contact the principal for details before scheduling repair work.

KEEPING A QUALITY MAINTENANCE CREW

Maintaining a quality maintenance crew is essential to implementing an effective school maintenance program. However, it is difficult to recruit and retain quality maintenance staff in public school districts because the market offers attractive salaries for comparable positions. Many maintenance staff prefer to stay with the school district because of benefits packages that can range from 20 to 30 percent of their base salary. It is not uncommon to see licensed maintenance staff picking up other jobs on the side after work hours to supplement their salaries. Some experienced maintenance staff leave the school district for an opportunity to start their own full-time business. To recruit and retain a quality school maintenance crew, the following approaches are recommended:

1. Salary. Work on the current salary scale of the maintenance staff to narrow the gap in salary between the school district and the market. Consideration should be given to initiating a merit pay system to award excellent work.

2. Rank. Create a rank system to indicate the position and rank a maintenance staff member holds. The rank system will allow the staff to see the opportunity for future career development.

3. Pride. Focus on work ethics and excellence by recognizing outstanding performance. Let maintenance staff take pride in what they are doing. Appreciation lunches have been proved to work.

4. Support. Support maintenance staff by securing resources for school maintenance. Build up staff morale by protecting maintenance budgets. Support maintenance staff by securing professional development opportunities.

5. Involvement. Involve maintenance staff in the operation of the maintenance department. Share with them the authority of management decisions. An involved team is a dedicated team.

ENERGY CONSERVATION PROGRAMS

New school buildings constructed in the last 20 years are largely energy conscious buildings. Most were built with a highly insulated shell and a limited amount of window space for energy conservation. The windows are made of thermal paint glass and all exterior doors are weather-stripped. They are installed with energy efficient boilers and HVAC controls. For buildings of this type, most school districts install a computerized programming system for the HVAC units that can be controlled at the maintenance department of the central offices. Efforts have been made to install the same systems in older school buildings so that eventually all school buildings will be under the same programming system for HVAC energy control. Many school districts continue to monitor the operation of the boilers at school buildings. Old, inefficient boilers are often replaced with energy efficient ones. Since the highest percentage of energy consumption is from lighting, most school districts have chosen to use standard, efficient fluorescent lighting for all schools and have worked with individual schools to develop plans to switch off lights when classrooms are not in use.

For energy conservation purposes, some school districts are willing to invest in improving the energy efficiency of old school buildings. Retrofit work in old school buildings includes the installation of HVAC controls, standard fluorescent lighting, efficient boilers, weather stripping, insulation wall panels, new ceilings, and new windows (Chan, 1980).

SUMMARY

The purpose of this chapter is to outline the different aspects of the organization of a school district maintenance department. It starts with a school maintenance program planning the budget development of a maintenance department. The daily operation of the maintenance department is then explored, followed by general guidelines recommended by the authors to prioritize maintenance requests and build a quality school maintenance team. The chapter ends by highlighting energy conservation projects often conducted by school district maintenance departments.

EXERCISES

1. Maintenance requests are received from schools and other administrative buildings. Do the requests from schools have priority over the requests from other locations? Why?

2. To maintain a school with high technology, the maintenance department could either send its staff for further training in technology or contract the technology work to outside technology firms. As director of maintenance, what direction would you prefer to take? Why?

3. Compare the current salary schedule for maintenance staff in your school district with those available in local businesses. What is your reaction?

4. As a new director of school maintenance, how do you organize an effective and efficient maintenance department?

5. Based upon the content of the district-level maintenance program, develop a school-level maintenance program to reflect the goals and objectives of the district-level program.

6. Use your school as a trial unit. Pilot test the school energy audit program of your school district. Write up a report to discuss your experiences in this involvement.

7. The school maintenance department is a large department with many divisions and shops. How do you determine the number of maintenance staff needed in each unit? Study the organization of your school district maintenance department and submit a justification.

8. Examine how computer technology is incorporated in the daily operation of the maintenance department of your school district. Can you recommend other uses of computers in school maintenance in addition to what has been done in your district?

5

SCHOOL BUILDING MAINTENANCE AT WORK

SCENARIO

The Brown County Middle School, the Brown County School District's newest facility, was constructed for $10 million with a capacity of 1,200 students. It represents the latest developments in modern architecture and technology. To protect the school district's huge investment, Mrs. Holly, the newly-appointed school principal, decides that a school building maintenance plan should be implemented from the inauguration of the new school. She disputes the idea that a new school building does not need much maintenance. She strongly believes that an effective maintenance plan starting from the first day of the new school building will protect the school building from rapid deterioration. As a result, the building could be used for an extended number of years. Mrs. Holly started by consulting with the school district janitorial consultant to recruit a capable head custodian. She then worked with the school design architect and engineers to offer orientation sessions of different building operating systems at the school. In each of these sessions, she made sure that district maintenance staff, the head custodian, the assistant principal, and other relevant faculty and staff were present to learn the operating systems of the new school building. She then organized a facility maintenance committee to plan for school maintenance. Mrs. Holly has been an educator for a long time, but school maintenance is definitely not her strength. She depends on her capable and reliable staff to give her the best advise. However, she is knowledgeable about the general planning guidelines. She has instructed the committee to include goals and objectives, activities, implementation procedures, and evaluation as basic components of the school maintenance plan.

UNDERSTANDING SCHOOL MAINTENANCE

A basic function of school maintenance is to keep the school building clean and safe so that it can be used to support educational functions. Cleanliness means not only emptying trash baskets and mopping floors but also maintaining sanitary conditions throughout the building, particularly restrooms and food preparation areas. Safety means that the school building has to be free of hazardous conditions that could be harmful to health. School buildings need to be checked periodically for fire safety, asbestos, radon, indoor air quality, and lead content in drinking water. At the same time, maintaining a pleasant school appearance is an essential aspect of school maintenance. Well-maintained lawns and attractive landscaping become the focus of public attention and shining floors and beautiful green plants inside the school brighten up the learning environment.

The tremendous effort of school maintenance cannot be done only at the school level; it requires district-level support. A collaboration of school and school district effort is needed to keep the maintenance program running. In fact, schools need to follow the maintenance policies and regulations developed at the school district level. School-level maintenance plans are developed to reflect and achieve the school district maintenance goals. Routine maintenance work at schools is performed by school custodial staff. All other repair work, such as plumbing, electrical, mechanical, doors, locks, and windows, etc., are the responsibility of the district office maintenance department. Maintenance requests are made by the school to the maintenance department for needed repairs.

In addition to maintenance by request, the maintenance department develops its preventive maintenance program by placing schools and their operating systems under a cycle of periodic checkups and services. The life expectancy of certain components of the operating systems are estimated and funds are budgeted yearly to cover the replacement costs of worn out parts and components as anticipated.

ROLES AND RESPONSIBILITIES OF A SCHOOL PRINCIPAL

Being the chief administrator of the school, the principal has overall responsibility for the school building. In school facilities, he or she is the facility leader who is expected to play a major role in developing an effective school maintenance plan. Specifically, the school principal is committed to creating a clean and safe school environment for teaching and learning. To accomplish this goal, the school principal leads, directs, supports, supervises, and evaluates the school maintenance plan. To initiate the maintenance planning process, the principal forms a facility committee responsible for developing the school maintenance plan. Committee members, consisting of the head custodian, faculty, staff, and parents, can be elected or appointed by the school principal. He or she assigns responsibilities to the committee and directs the development of the maintenance plan. As facility leader, he or she supports and implements the plan with available resources. An important aspect of the principal's facility responsibility is the supervision of school custodians. He or she assigns daily maintenance responsibilities to the head custodian and works with the head custodian to supervise the work of other custodial staff. The principal evaluates the custodians' performance according to preset criteria. He or she also evaluates the outcome of the school maintenance plan by measuring noticeable progress and outcomes.

ROLES AND RESPONSIBILITIES OF A CUSTODIAN

The custodial staff in a school bears the most important responsibility for maintaining the school building as a clean and safe physical environment to support teaching and learning activities. The staff consists of a head custodian and custodians (the number to be determined by either the school student population or building square footage per school district formula). The head custodian assumes many roles: a planner, a supervisor, a communicator, a worker, a coordinator, and a leader. Under the direction of the school principal, the head custodian plans to implement the school maintenance plan as determined by the school facility committee. He or she needs to perform the following tasks:

1. Identify the major aspects of the maintenance plan;
2. Determine the staffing needs to complete the task;
3. Recruit custodians per school district allotment;
4. Check on availability of equipment and supplies and purchase as needed;
5. Be familiar with the various building systems and their controls;
6. Divide maintenance work among the custodians by building section and by personal skill;
7. Plan custodial work schedules;
8. Supervise the quality of work assignments;
9. Work with custodians to improve efficiency and effectiveness;
10. Check on the daily operation of the school building systems;
11. Repair minor damage to school buildings;
12. Submit requests for any maintenance work beyond the school level;
13. Plan for any long-term maintenance work during the holidays;
14. Record and report periodically on the physical conditions of the school building;
15. Keep proper records of maintenance work performed at the school;
16. Perform evaluations of custodians on an annual or bi-annual basis; and
17. Assist faculty and staff in their daily duties as needed.

Custodians work as hard as the head custodian in their areas of assigned responsibilities. Because of the nature of their work, custodians may work different shifts to get the work completed. All custodians need to recognize and be recognized that they play an essential role in the operation of school facilities.

DEVELOPING A SCHOOL MAINTENANCE PLAN

In developing a school maintenance plan, the school facility committee needs to work toward reflecting the goals and objectives of the district-level maintenance program. Since every school building is constructed differently, the time and effort required to maintain school buildings vary according to the particular design features installed in the school. Therefore, the first step in developing a school maintenance plan is to conduct a school maintenance audit that includes:

1. An inventory of the characteristic features of the school building;
2. What needs to be done to properly maintain the building for educational purpose; and
3. How many man-hours and what types of equipment and supplies are needed to complete the anticipated maintenance work.

In most school districts, school maintenance responsibilities are shared between the school custodial staff and the school district maintenance staff. All the school maintenance items on the audit list are divided between the school staff and the district staff. These maintenance items, described by Schaefer (1967) as operational activities, include housekeeping (cleaning and sanitation), operation of the mechanical and electrical plant, safety and security, equipment servicing, upkeep of grounds, minor repairs, and occupant services. With all the basic information in hand, the budgeted allotment for custodial staff and maintenance equipment and supplies is examined. The time element has to be considered in determining individual workload. Some maintenance items need to be performed daily and some periodically. Some items take much longer time to complete than others. As a result of the school maintenance planning, an annual schedule of maintenance events is developed. These events include daily and periodic operating items by custodial staff and school district maintenance staff. In addition, the schedule should include projects the school district preschedules to be implemented in a maintenance cycle, such as reroofing, repainting, carpet replacement, boiler replacement, etc. An implementation component of the school maintenance plan should include supervision, operational decisions, and budget control functions. The plan should also have a built-in evaluation component to enforce interim and end-of-year evaluations based on the goals and objects established in the development of the maintenance plan.

THE PARTICIPATORY CONCEPT IN SCHOOL MAINTENANCE

School maintenance activities so far described have been participatory in nature. Examples of the participatory concept are shared responsibility between school custodial staff and school district maintenance staff and the development of the school maintenance plan by a diversified school facility committee. The participatory concept of school maintenance can be further utilized in checking for school building safety. A checklist of school building safety can be developed to include all interior and exterior items. Members of the school facility committee can take turns to check on the school building safety items by using the checklist for evaluation. The use of this participatory approach to school maintenance serves three purposes: 1) it helps the school custodians tremendously in the time and effort they have to spend on tracking down the condition of the items on the list; 2) checking on school safety items can be done more frequently with assistance from more people; and 3) involvement of a diversified group of participants helps promote a sense of belonging among school building occupants. A sample checklist is included in appendix B.

STAFFING FOR SCHOOL MAINTENANCE

A school needs to be appropriately staffed to maintain functions without which the school cannot keep its standard of a clean and safe environment for learning. In staffing for maintenance, school principals have experienced several difficulties:

1. Salaries offered to custodians are relatively low for the qualifications the schools call for;
2. No preparatory program or certification is offered for potential custodians; and
3. Limited opportunity for career development is offered to custodians.

As a result, school principals find it difficult to hire and retain qualified custodians to work at schools.

Hiring

Custodial positions are not usually publicly advertised. Position announcements and job descriptions are filed and posted in schools and in district central offices. Principals searching for custodians look up the files available at the personnel department of the school district central office and make personal contacts. The search process is simple. Principals, together with the assistant principals and the head custodian, review the applicants' qualifications, conduct the interviews, and make hiring decisions. The National Center for Educational Statistics and the Association of School Business Officials International (2003) have a model job description for school custodians that includes essential duties and responsibilities, including daily duties, weekly duties, monthly duties, winter and spring break duties, summer duties, working conditions, equipment used, general qualifications, educational requirements, credentials, licenses, and physical requirements. The specific descriptions can be modified in searching for custodians with particular skills. "Custodians should exhibit a sense of cooperation, an ability to work with others, and a respect for teachers and students" (Kowalski, 2002, p. 242). In searching for a head custodian, in addition to the basic qualifications required of a custodian, emphasis is placed on personal characteristics, leadership skills, supervisory skills, technical skills, and past experience. Applicants with excellent records of previous experience as custodians in other educational settings are preferable. The maintenance department in the school district central office provides a good source of references for the candidates.

Training

When custodians are hired to work in a school, they need to go through an orientation process. They need to be introduced to the school organizational structure and be guided to understand their importance in the organization. Getting to know their colleagues, and their specific assignments, is essential to their new positions. All custodians need to be familiar with the physical features and design of the school building they serve. The head custodian needs to learn the operational procedures of the building systems and their controls. In this respect, staff in the maintenance department of the central office is most helpful to work with school custodians to make sure that they possess the skills to correctly operate the building systems. All custodians need to fully understand the extent of their work in maintaining a school building. They need to call the maintenance department for assistance with the types of work they are not licensed to do. Many new custodians need to be trained or retrained for new skills required for school operation. For beginning custodians, the maintenance department of the school district offers janitorial workshops from time to time to get them acquainted with their new position. The head custodian also has the responsibility of training newly hired custodians to use new equipment and supplies.

Supervision

It is the responsibility of the school principal to supervise the custodial work to ensure a clean and safe school environment for learning. Sometimes the principal may choose to dedicate the responsibility to the assistant principal, who will update him or her on a daily basis. The principal or assistant principal supervises the head custodian and the custodians. The head custodian in turn supervises all the custodians. The scope of supervision is to check and make sure that the daily, weekly, and monthly custodial assignments are satisfactorily performed according to professional standards. It must be emphasized that supervision of custodial work has to be taken very seriously because school administrators cannot afford charges of negligence leading to unhealthy and unsafe school buildings. Custodians are expected to understand the outcomes of their work and to perform in a manner to achieve the outcomes. All good work needs to be confirmed and complimented. When a custodian is not performing to expectation, he or she will need to be closely supervised. Assistance can be offered by assigning him or her to work directly with the head custodian or another custodian with good performance. He or she may also be retrained by specialists of the maintenance department in districtwide janitorial workshops. Principals may try the team concept of custodial work organization, by which custodians of different special skills are assigned to form a working team. In this way, each custodian is given the opportunity to perform in an area of his or her expertise. The result is achieving the best performance by utilizing the best of custodians' skills.

Evaluation

The performance of custodians needs to be formally evaluated by the school principal or assistant principal and the head custodian at least twice a year in a formative evaluation and a summative evaluation. The purpose of evaluation is recognition of good performance and improvement of substandard performance. However, performance evaluation is used by some school districts as a means to determine merit pay. In the evaluation process, the outcome of custodial work is measured by using a set of predetermined criteria. Results of the evaluation need to be shared with the custodians in discussions on strengths and areas that need improvement. Custodians with below average performance need to be offered opportunities and assistance for improvement. The focus of the summative evaluation is the establishment of goals to work on difficult areas in the formative evaluation. The head custodian is evaluated by the principal and the assistant principal in the same manner. The principal or the assistant principal will work with the head custodian on areas that need improvement. All documents on evaluation should be confidentially filed in personnel records.

Staff Retention

A combination of reasons, such as low salary and low self-esteem, contributes to job instability for school custodians. To retain good custodians in schools, attention has to be paid to reworking the salary schedules and benefits packages of school custodians. A study of custodians' remuneration in the neighboring school districts could generate meaningful data to help make decisions to adjust the custodians' salaries and benefits. Competitive salaries and benefits are powerful tools to retain good custodians. Some school districts have designed a rank system of custodial positions to include janitorial specialist at the district level, and head custodian, custodian class I, and custodian class II at the school level. The rank system offers school custodians more opportunities to move up professionally in the same school district. In addition, strong support from administration in terms of resources makes it easier for custodians to satisfactorily perform their daily duties. School principals need to work with custodians in recognizing their significant contributions to schools. Special occasions, such as honor's day, appreciation day, Thanksgiving dinner, Christmas party, or dedication ceremony, offer unique opportunities to publicly thank the custodians for their great work. Custodians with higher levels of self-esteem tend to have better performance and greater job stability.

SCHOOL MAINTENANCE EQUIPMENT AND SUPPLIES

The need for school maintenance equipment and supplies is derived from the special design features of a school building. This is related to the school facility inventory discussed previously. Information from the facility inventory will determine the most efficient and effective maintenance procedures. Then a list of equipment and supplies can be prepared accordingly to facilitate the maintenance process. For example, to determine what equipment and supplies need to be used to maintain the floor covering of a school, the specific types of floor covering together with corresponding square footage should be calculated. Product specifications and maintenance instructions from the manufacturers of floor covering materials should be obtained. The qualities of the different floor covering materials should be studied in detail to decide on the maintenance procedures and schedules, and the equipment and supplies. Janitorial specialists from the school district maintenance department and consultants from product manufacturing companies can assist in recommending the selection of maintenance equipment and supplies. However, the selected equipment and supplies need to be tested to verify they work. Sometimes, small quantities of several types of supplies have to be tried to determine which one produces the best effect before large quantities are stocked.

SUMMARY

This chapter highlights the major aspects of facility maintenance at the school level. It starts with the development of a school facility maintenance plan. The school principal and the head custodian play major roles in the implementation of the plan. Their specific responsibilities in the maintenance process are discussed. Since school maintenance is a joint venture between the school and the district maintenance department, discussions are held on the balance of work and the line of authority to be followed. Staffing is a main issue in school maintenance, and how to recruit, train, supervise, evaluate, and retain good school custodians is fully explored. The chapter ends by suggesting an approach in determining equipment and supplies for school maintenance. School maintenance is an investment for the future. It serves to extend the useful life of a school building and saves money for the school district in the long run.

EXERCISES

1. School restrooms need to be kept clean and healthy at all times. Take inventory of all the special features of restrooms in your school. Develop a

maintenance schedule for all restrooms and determine what equipment and supplies you plan to use.

2. Prepare a list of questions you would ask the candidates for the school head custodian position in an employment interview.

3. As school principal, develop a plan that works toward bringing out the best potential of your custodians.

4. The school district has developed a maintenance plan that covers all school facilities in the school district. Why do individual schools have their own maintenance plan? Is there any duplication of effort?

5. Interview a school head custodian and the director of maintenance of a school district. Find out how maintenance work of a school building is divided between the school and the district maintenance staff.

6. Write up an orientation plan for beginning school custodians.

7. Is there any difference in maintenance work among elementary school, middle school, and high school buildings?

8. Evaluate the custodial schedule of your school. What are the assigned responsibilities of each of the custodians? Develop an alternative schedule that focuses on improving efficiency and effectiveness.

III

SCHOOL RENOVATION

6

PREPARING FOR SCHOOL RENOVATION

SCENARIO

The Waterford School District has six elementary schools, two middle schools, and one high school. The current superintendent was elected on two previous occasions but is likely to be replaced when the board begins superintendent selection in the coming summer. The superintendent could best be described as a "good old boy" who graduated from the local high school, went to the nearby teacher's college, and eventually received administrative endorsements from the regional university. He has been under continual attack for the past two years for his perceived lack of leadership in addressing the needs of the school district. Parents complain that they want their children to be able to go to the state university and compete for scholarships and awards. The parents perceive that Waterford's curriculum is outdated and fails to address the needs of contemporary students. In fact, the parents were outraged recently when the superintendent advocated eliminating computer labs from the schools in an attempt to balance next year's budget. A recent issue confronting the superintendent is that certain local communities are pushing to revise the facilities planning priority by bringing Wilton Springs Elementary School and Doe Creek Middle School renovation projects ahead of the new high school construction. Community members are trying to justify their request by data collected from comparing facility needs from school to school. They have already signed up for a presentation at the upcoming school board meeting. To prepare for a response to the community at the school board meeting, the superintendent has asked the director of school planning to send him information about the two schools.

Wilton Springs Elementary School

Wilton Springs Elementary School presently houses 875 students in grades kindergarten through four; however, the school was originally designed for only 650 students. As a result, there are 12 mobile classrooms on the campus to house regular classes in addition to special needs classrooms (e.g., special education, gifted education, Title I, etc.). Wilton Springs has one principal, one assistant principal, one guidance counselor, one media specialist, two school secretaries, eight paraprofessionals, and a faculty of 49, including special needs teachers. The school is located on a 12-acre tract of land adjacent to Doe Creek Middle School. It is a 25-year-old building with typical brick/block construction arranged in pods. Originally designed as an open-space school, walls have long since replaced the open spaces between classrooms. The Wilton Springs community is largely blue collar with a few white-collar (small business type) jobs in the area. Four large industrial plants relocated to the Wilton Springs area during the past decade and now dominate both the employment and the political agenda of the area. Approximately 45 percent of the students at Wilton Springs are African American, 50 percent are white, 4 percent are Asian American, and 1 percent are Hispanic.

Doe Creek Middle School

Doe Creek Middle School occupies the "old" Wilton Springs High School building originally constructed in 1954. The building underwent major renovation during the late 1960s and again in 1981. Currently it has 1,325 students in grades five through eight, although it was originally projected for a maximum of 1,200 students. Doe Creek has one principal, two assistant principals, three guidance counselors, six paraprofessionals, two media specialists, two school secretaries, one bookkeeper, one faculty clerk, and a faculty of 57, including special needs personnel. When the school was opened originally, the principal and faculty were selected from the "old" high school and junior high school and did not understand the middle school concept. However, with the nurturing effort of three principals and numerous teachers, the middle school concept has been truly accepted and implemented in Doe Creek Middle. Approximately 40 percent of the students are African American, 53 percent are Caucasian, 5 percent are Asian American, and 2 percent are Hispanic. In addition to the request to move the two renovation projects ahead of the original schedule, community members also asked the school board to examine the overcrowding conditions in the schools and consider possible additions to existing school buildings.

WHAT IS SCHOOL RENOVATION?

School renovation is work planned and constructed to change the physical arrangement or appearance of a school facility (Kowalski, 2002). Typically, school renovation is designed for the school building to better function and support educational and instructional activities (Aller, 2002). Renovation may take the form of simple remodeling or a total makeover (Argon, 1997). Renovation will change the design and configuration of the building and will change the instructional atmosphere of the building (De Hann, 2000). Typically, the number of instructional areas will be changed because of renovation (DiBella & Anderson, 2000). The purposes of a school renovation project are to satisfy the educational program needs and to uphold the building structure safety standards.

ASSESSING SCHOOL FACILITY NEEDS

School buildings continue to get older and revenues for constructing new facilities continue to decline (Argon, 1998). An average school in the United States is estimated to be almost 50 years old (Honeyman, 1998b). Student enrollment is constantly shifting as the overall population shifts for personal or economic reasons (Kowalski, 2002). Consequently, school administrators are faced with the dilemma of building new facilities or trying to be efficient and renovate rather than build new facilities (Hawkins & Lilley, 1998).

Much of the declining condition of schools can be attributed to deferring school maintenance due to unavailable funds (Hartman, 1988; Jarvis, Gentry, & Stephens, 1967; Wood, 1986). Deferring maintenance can be cost effective for the short term, but catastrophic for the long term (Jarvis, Gentry, & Stephens, 1967; Lackney, 1999; U.S. Department of Education, 2000; U.S. General Accounting Office, 1996). Short-term problems have a tendency to become long-term problems very quickly in schools due to their heavy use and the unpredictable nature of people using the facilities (Rivera-Batiz & Marti, 1995).

School administrators should conduct a comprehensive inventory of current educational facilities and compare it to the future needs of the school district to develop a facilities plan that projects the needs of the district for at least 10 years (Kowalski, 2002). Basically, there are four ways of obtaining the data for the facilities plan, any of which will provide the needed information. First, the district could commission a private consulting firm that has expertise in school facility management (Swanson & King, 1997). Second, the district could contract with the state department of education or other educational agency (i.e., local university) to conduct the needs assessment for facilities, often at little or no cost to the district (Thompson, Wood, & Honeyman, 1994). Third, the school district could conduct the needs assessment using school district personnel, particularly those skilled in evaluation techniques (Kowalski, 1989). Fourth, the school district could select some combination of the other three methods. Regardless of the method chosen by the school district, the outcome should be the same: a workable and manageable long-range school facility plan (Duke, Griesdorn, Gillespie, & Tuttle, 1998).

The school facilities planning model in figure 6.1 is a classic example of how school districts commonly pursue their effort to achieve long-range planning of school facilities.

This model illustrates the need for the school district to examine the current facilities that are available and compare that to an ideal facilities plan. The difference between the ideal and the real constitutes a workable and manageable facilities plan that can be utilized by the school district to meet the educational needs of students.

The facilities plan should incorporate the following:

1. Present education facilities;
2. Educational goals and objectives;

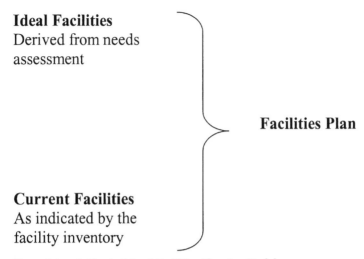

Ideal Facilities
Derived from needs
assessment

Facilities Plan

Current Facilities
As indicated by the
facility inventory

Figure 6.1.   A Classic School Facilities Planning Model

3. Community survey (the needs of the community);
4. School population trends;
5. Current educational programs;
6. Quality of educational facilities;
7. Quantity of educational facilities;
8. Economic aspects of capital outlay;
9. Infrastructure resources; and
10. Current and future planning (see Abramson, 2002; Aller, 2002; Carter, 2003; Flanigan, Richardson, & Stollar, 1995; Hansen, 1992; Kowalski, 2002; Thompson & Wood, 1998).

THE NEED FOR SCHOOL RENOVATION

What are the reasons for school renovation? The most obvious reason is that renovation is more economically marketable in today's tight money economy. Typically, the reasons can be classified as follows:

1. Rapid student growth or change;
2. Implementing federal or state regulations;

3. Sustained growth;
4. Planned change and modification;
5. Renewing space and facilities;
6. Implementing new needs (technology);
7. Applying standards for instructional needs (small class size);
8. Community expectations and demands;
9. Utilization of existing facilities (renovate old high school for middle school); and
10. New configurations (magnet schools) (see Aller, 2002; Argon, 2001; DiBella & Anderson, 2000; Earthman, 1998; Gisolfi, 1999; Hoskens, 2003; U.S. Department of Education, 2000).

The core concept of school renovation is to keep existing facilities for a longer time so that the school district's resources can be more advantageously utilized. Obviously, the consideration is how much longer the facility will last after renovation and how much money is involved in renovation to be justifiable. Without doubt, decisions on school renovation have to tie in with an architectural evaluation of the school building to examine how much it will take to bring the building systems back into shape. Then there is an educational evaluation to find out if the building could adequately serve its intended educational program after renovation. Comparison has to be made between the costs of renovation and the costs of new construction before a renovation decision is made.

PLANNING FOR SCHOOL RENOVATION

Most renovation planning can also be called contingency planning (Cook & Hunsaker, 2001), for example, the belief is that there is no one best way to think about the facilities. That is not to say that any way is as good as any other; the concept is that different schools exist in different conditions and face different problems (Boyles, 1995; Smith & Ruhl-Smith, 2003). Therefore, administrators need to think about those conditions and adapt their planning and administrative style to them (Davis & Tyson, 2003).

Administrators who conduct facility planning should also remember that planning is difficult for a variety of reasons. Most often people resist planning because they perceive that it will lead to change and many educators are reluctant to change (Cook & Hunsaker, 2001). Planning efforts should be aimed at clarifying the opportunities and obstacles facing the facilities plan and should focus on the major needs of the school district (DeJong & Glover, 2003). Planning is also a learning process for the many people who will be involved in the process (Richardson & Lane, 1997). The learning will be about the educational program, the people in the program, and the facilities that accommodate them. Therefore, planning will improve communication in the district as more people contribute to the planning effort. The knowledge explosion has forced more organizational planning because no one individual or group can provide or contain all the information needed to make quality decisions (Cuban, Kirkpatrick, & Peck, 2001). Planning must be viewed as a continuous process, not an event (Richardson, Short, & Lane, 1996). By involving numerous people in the planning process, the district will encourage cooperation, not competition, for scarce resources (Richardson & Lane, 1997). Finally, the planning process should help develop a sense of ownership among the participants, who are more likely to vocalize their support for the facility plan (Kosar, 2002; Richardson, Short, & Lane, 1996; Sanoff, 2002).

SCHOOL RENOVATION AS A COMPONENT OF THE SYSTEM FACILITIES PLAN

Every school district should develop a master plan for the modernization or replacement of existing facilities, coupled with the need for new structures (Kowalski, 2002). This plan should be developed in consultation with community members and other stakeholders (Cecil & Boynton, 1999; Connor, 1998; Davis & Tyson, 2003; Henry, 2000; Kosar, 2002). The plan should be of sufficient quality to provide the direction for future growth, sustainment, or retrenchment (Hansen, 1992; U.S. Department of Education, 2000). The plan should reflect several items, including:

1. Educational goals
2. Feasibility
3. Supporting resources (existing infrastructure, e.g., streets, utilities, etc.)
4. Enrollment patterns
5. Building conditions versus needs
6. Transportation patterns

7. Projected costs for new construction versus renovation
8. Projected costs for bonds versus pay-as-you-go
9. Safety
10. Educational adequacy
11. Location adequacy
12. Site adequacy

School renovation is an essential component of this system-wide facilities plan. Many school buildings can be seriously considered for possible renovation. Renovation saves money by saving the system from the expense of building new facilities. In addition, renovation projects have to be prioritized with reference to the growth demographics and needs of different areas within the school district.

THE SCOPE OF RENOVATION PROJECTS

The decision to renovate an existing facility versus building a new facility should not be made lightly or in the heat of the moment. Renovation requires thorough and careful planning. Renovation should not be about moving one wall for another classroom but rather about the vigilant and deliberate changing of the facility to accommodate educational needs (Lackney, 1999).

Persistent pressure groups, most often with their own special agenda, habitually attempt to influence the direction of renovation. These pressures can lead to economic waste, coupled with educational and facility inequities in the same district. Community pressure groups always want what they perceive to be best for "their" school and not what is best for the school district (Duke, Griesdorn, Gillespie, & Tuttle, 1998).

Many aging schools have difficulty with energy efficiency and other conditions that could affect student safety (Sturgeon, 2000). Additional features of school safety include the installation of restraining materials (fences, etc.) and examination devices (metal detectors, etc.) that are often problematic for older buildings (Gottwalt, 2003).

In determining the scope of a school renovation project, the following factors need to be carefully considered: 1) accommodation of anticipated educational programs; 2) school building safety; and 3) school building efficiency. Most renovation projects include face-lifting items in the prescribed work.

Though face-lifting gives the school building an attractive look, it also needs to make room for essential considerations of building safety, efficiency, and educational functions.

MEETING FEDERAL AND STATE MANDATES

All schools are faced with the requirements of meeting state and federal mandates, which are often unfunded (Honeyman, 1994). The legislative branch of government can pass two types of legislation: authorization and appropriation (Alexander & Salmon, 1995). State legislatures have learned from Congress to authorize a program or a project and then not pass the required appropriation legislation to fund the program (Swanson & King, 1997). Rather, the authorization is "passed" and the requirements for appropriation are "passed on" to the state or local entity that is required to implement the program. During the past 20 years, several federal programs have had a tremendous influence on the facilities of local school systems (Burrup, Brimley, & Garfield, 1996; Wood, 1986). Asbestos abatement and The Americans with Disabilities Act of 1990 required schools to be "clean" and to accommodate the special needs of those using schools by installing access ramps, elevators, modified bathrooms, and expanded entrances (Thompson & Wood, 1998). Some school systems struggled with the removal of lead paint and radon gas (Honeyman, 1998b).

Almost all states have experienced some form of educational reform during the past couple of decades, from the total restructuring of the Kentucky system of education to states with few reforms (Ritter & Lucas, 2003). New requirements call for reduced class size and mandated modifications to accommodate different teaching and learning options for all affected students. Such accommodations have also encroached on educational facility development, particularly since much of the funding in education goes to accommodate programs and very little has been appropriated for facilities (U.S. Department of Education, 2000). In 1999, the federal government attempted to change that procedure and provide funds for some school districts to request based upon need (U.S. Department of Education, 1999). Some schools need small-group and large-group facilities, in addition to various needs for multimedia resource centers (Honeyman, 1990). Other needs include computer labs and classroom labs for science and mathematics (Cuban, Kirkpatrick, & Peck, 2001). In addition, the federal No Child Left Behind Act has forced school administrators to critically examine how school facilities might affect student achievement, as accountability has become the current buzzword in education (Ritter & Lucas, 2003).

TECHNOLOGY MANDATES

Advances in technology have created a facility nightmare for many school administrators, who need to make changes to almost all school environments to incorporate new technology (Bradley & Protheroe, 2003). Many schools lack conduits for computer cables. Many more lack the electrical capability to handle the computer infrastructure (e.g., electrical outlets, wiring for technology, etc.). Technology has the potential to connect schools with the world, which could be a major boost to student progress. However, unless the technology infrastructure is present, students cannot realize the full potential of technology (Rittner-Heir, 2003).

With advanced technology, teachers have additional resources to reach more students in new and innovative ways. To take advantage of these resources, buildings, old and new, must be capable of supplying the technologies to the classrooms (Coley, Cradler, & Engle, 1997). These buildings are the places where teachers prepare the younger generation for the rest of their lives. Wiring can become part of the infrastructure in new buildings (Hawkins & Lilley, 1998). However, the vast majority of schools in this nation are not new and wiring an old building often provides significant challenges for any school system (Butterfield, 1999). The job of retrofitting older facilities can be a daunting and expensive task (Aller, 2002; Flanigan, Richardson, & Stollar, 1995; Thompson & Wood, 1998).

Educators are teaching children how to live successfully in their future. Today, teachers are still asked to teach students to think, but instead of imparting knowledge, they now must help students understand where and how to find knowledge due to the proliferation of information evident in this technological world (Richardson, Chan, & Lane, 2000). Finding information is as important, or more important, than knowing facts that may be obsolete in a short while (Drucker, 1999). Exposing students to the wealth of information sources becomes the job of the teacher, and that knowledge is available only through the technology of today's world (Coley, Cradler, & Engel, 1997). It then becomes critical for teachers to understand the technology, but more important that school administrators provide the hardware and software to the teacher to aid the student (Cuban, Kirkpatrick, & Peck, 2001). Many systems and schools operate from crisis to crisis and do very little long-range planning and goal setting (Richardson, Short, & Lane, 1996). Certainly one of the goals of every system and school must be the preparation of students for society, a society that will be far different from today (Nixon, 1998). To reach that goal demands planning for instruction and for technology. The rapid introduction of technology demands that teachers know and be able to use technology (Means & Olson, 1995). Yet, that introduction is meeting resistance because some schools and school systems do not have the infrastructure in place to provide the needed technology (Cuban, Kirkpatrick, & Peck, 2001). Technology can be a positive for everyone in the school system, but it requires planning because the world is on the fast-track of the information superhighway and education has to be an active participant (Richardson, Short, & Lane, 1996).

BUDGETING FOR SCHOOL RENOVATION

Budgeting is the means through which educational planning becomes reality (Flanigan, Richardson, & Stollar, 1995). The budget represents educational needs in monetary terms that are understood by both professional educators and the lay public. School renovation budgeting is most often categorized as a part of capital outlay budgeting. According to Flanigan, Richardson, and Stollar: "The first step in developing a capital outlay budget is a comprehensive survey of present educational facilities in relation to present and future educational goals and objectives" (1995, p. 184). Thereon, specific renovation projects can be identified per evaluation of individual buildings. Building evaluation is an important step before actual budgeting begins because not all school buildings are worth renovating. In a school building identified for renovation, a priority list needs to be developed of all items for possible inclusion in the project. Then an estimated price tag can be tied to each of the items. In most cases, school districts cannot afford to complete all the desired items. Consideration has to be given to items of high priority and determination has to be made to bring the project to a reasonable and affordable scope. Because of hidden conditions in renovation projects, a sizable amount of funding has to be reserved for contingency use. Fifteen to 20 percent of the baseline budget is recommended for contingency purposes.

SUMMARY

American education has a long and distinguished history. The early settlers viewed education as a function of the church. It was nearly 100 years later that a local entity taxed its citizens for educational purposes. The framers of the U.S. Constitution did not provide for a national role in education; rather, they reserved that function for the individual states. As a result, there are 50 different educational systems in the United States. These different systems have created both challenges and unique opportunities in American education. One of the major concerns in all American education is how to fund schools.

As states have attempted to provide for the increased needs of their clientele, locating additional funding sources has presented major difficulties for educational administrators. American citizens want more from education, but they are reluctant to pay for the services. Only one in three Americans own property, yet most local school districts are funded by local property taxes, Michigan being the notable exception. Only one in four Americans has children in school, yet everyone benefits from the education of children who become productive citizens.

In recent years, much has been written about the disrepair of America's schools, yet school administrators continue to search for more funding to replace or repair school buildings that were built approximately 50 years ago. Building or renovating schools is a costly proposition, yet effective instructional methods demand facilities that are conducive to learning. What is the answer to these dilemmas? There is not one answer; educational administrators are faced with changes that have severely interfered with the normal expansion of educational facilities to meet the changing demands of Americans for their educational institutions.

EXERCISES

1. Visit a school that is being considered for renovation. Request the following information:

   A. What is the driving force behind the renovation?
   B. Is this a total renovation or a partial renovation? Why?

   C. What is the age of the building?

   D. What is the history of the building?

2. Conduct a needs assessment for the school.

   A. What are the major educational needs?

   B. Will this renovation meet those needs?

3. Interview the architect who is responsible for the renovation.

   A. What is the architect's vision for the renovation?

   B. Who participated in planning the renovation?

   C. How will the renovation be evaluated?

4. Develop a list of school building items that could be included in a renovation project. Prioritize all the items on the list and provide your justifications.

PLANNING AND DESIGNING SCHOOL RENOVATION

SCENARIO

B atesville County High School is housed in a 40-year-old building. Because of growth, the school district has placed 15 portable classrooms on campus to meet the need for classroom space. The decision was made two years ago to build a new Batesville County High School that would be large enough to accommodate anticipated growth in the next five years. The decision was also made to renovate the old high school into a middle school. While the new high school is under construction, with a completion date one year from now, planning is ready to begin for the renovation of the old high school building. The director of planning in the district office is facing a number of issues in this renovation project. First, the public is eager to preserve the historic look of the old high school building as a symbol of the community. Second, substantial interior redesigning is needed to meet the requirements of the current and future middle school program. Third, the state, by renovation formulae, has allocated $1 million for the renovation project. But more construction dollars are needed for the amount of renovation work already called for. Therefore, in the last school board meeting, the board members voted in support of the renovation by authorizing an $800,000 matching fund from the local school district and made it clear that this was all they would approve. Fourth, the local fire marshal and the building inspectors, knowing that the old high school building will be undergoing major renovation, have stipulated that the old high school building needs to be brought up to the current building and fire codes. The director of planning has to meet the challenges of satisfying the public's expectations, the educational needs, and the current fire and building codes with $1.8 million. With the increasing cost of construction, $1.8 million is not going to go very far in the renovation project.

JUSTIFICATIONS FOR RENOVATION

In planning for school renovation, a committee needs to be organized consisting of school administrators, teachers, administrative staff, the head custodian, media specialists, the cafeteria manager, and community leaders. The school's district central office should be represented by the maintenance director, the planning director, the curriculum director, the purchasing agent, and the transportation director. The planning committee needs to examine the purpose of the renovation. In the scenario, the main purpose of renovating the old high school is to convert it to be used for a middle school. Obviously, the facility requirements for accommodating a middle school program are different from accommodating a high school program. The academic organization of a middle school is by grade and by team, whereas the academic organization of a high school is by department. Middle schools are structured in school-within-school clusters. High schools are designed by areas to serve academic departments. Understanding the basic reason for the renovation is essential in developing the school renovation plan. Anything included in the renovation needs to be related to the purpose of the renovation project. Then educational specifications can be developed to indicate the educational goals, objectives, activities, and space requirements of each academic area of the middle school. Educational specifications serve as guidelines to the architect and the engineers to redesign an old school building to fit its new use.

Another justification for renovation is aesthetics. Renovation work addressing building appearance may not be directly related to educational purpose, but it is expected of a renovation project. After all, a renovated school building has to look attractive. Some of the justifications also come from a safety standpoint. Any problem with the structure, roof, and electrical, mechanical, and plumbing systems of the school building have to be addressed. Also, meeting new fire and building codes is another strong justification for renovation.

THE EXTENT OF SCHOOL RENOVATION

An old school building is renovated not only to serve the new educational purpose but also to last as an educational facility. If money is not an issue, then renovation work can be performed in many different ways to satisfy all needs. However, with a limited budget, the constraint is to address the priority concerns first. In deciding on the extent of school renovation, the planning committee can list all the work that needs to be done in order of priority. The list has to stop where funding would be exhausted. The reality is that the funded amount for school renovation will never satisfy the demand for work that needs to be done. Understanding the nature of school renovation, the planning committee can start working on the list of items that could be included in the renovation work. The architect and the engineers can help identify major items of safety concern. The committee as a whole will decide on the priorities. Some of the renovation items are nice to have if funding is available. A common practice is to place these items on the list of alternatives in the bidding process. This allows the school district the flexibility of accepting or rejecting the alternatives. Priority concerns on a school renovation list include: safety, code updates, educational program requirements, and aesthetics.

Safety

Safety is the primary concern for planners in a school renovation project. The project architect and the engineers will check to ensure that the building is safe and can continue to function as a school building. Any structural failures detected have to be addressed as the number one priority. Some of the common safety problems found in old school buildings are roof leaks, excessive moisture, foundation saturation, construction materials containing asbestos, wall cracks due to excessive expansion and contraction, poor indoor air quality, faulty wiring, rusted pipes, and termite infiltration.

Code Updates

Many old school building designs no longer meet today's fire and building codes. Since the building was approved when it was first built, the school district could continue to use the building without having to change anything every time the codes are updated. However, in most renovation cases, the local building department and fire department will require the school district to bring the renovated school building up to the current codes.

Educational Program Requirements

The planning committee will try its best to meet the requirements of the new educational program as outlined in the educational specifications. In most cases, classrooms may need to be enlarged, the spatial relationship between rooms may need to be reworked, some rooms may need to be relocated, and some room functions may need to be changed. Because of the structural design of the old school building, changes requested to suit the new program may not be possible without extensive alterations.

Aesthetics

School building aesthetics serve an essential educational function in addition to enhancing public relations (Chan, 1988a). The outcome of any school renovation project is the improvement of the building's look. The community wants to see where its tax dollars are spent and aesthetic items are the most visible result of renovation. Therefore, most renovation work in schools includes new wall paint, new floor carpet or tiles, new ceilings, new lighting, new restroom partitions and fixtures, and new cabinets.

Other renovation items may include new furniture and equipment, new school signage, resurfacing or repaving drives and parking areas, landscaping, new doors and locks, new windows, new HVAC units, new boilers, and energy conservation devices. The list can be very long, but the planning committee has to stop at a certain point when the budgeted amount is exhausted.

SCHEDULING SCHOOL RENOVATION

Scheduling school renovation may take many different formats. Some school districts simply house the students in another facility when renovation work goes on in the school building and wait until the renovation work is complete before moving the students back in. This is the preferred arrangement because it will remove the risk of liability caused by placing students inside a building undergoing renovation. The disadvantage is the tremendous effort of moving the entire school setting in and out. Many school districts do not have the luxury of another vacant facility to house the students in while renovation goes on in the school. In this case, renovation is performed by stages. Portable classrooms can be hauled in to temporarily accommodate some students so that a section of the school building can be blocked off for renovation. When this section is completed, the students in the portable classrooms can move back in and then students in another section can move to the portable classrooms, leaving another section of the school vacant for renovation. This is not a preferred format, but in some situations it may be the only choice. The format of renovation by stages is expensive because of the extensive time involved. School administrators will have to work closely with the contractors to minimize disruption to school activities during renovation. Another format for school renovation is to schedule the work for the summer. Arrangements can be made to either end the current school year early or start the next school year late to allow more time for the renovation work to be completed in the summer. Extensive renovation work can be divided into two parts to be performed in two summers. However, arrangements should be made to get as much as possible of the renovation work completed in one summer to avoid further disruption to the school.

No matter what format a school renovation project takes, arrangements should be made to ensure the safety of building occupants and to cause the least disturbance to school activities. All stakeholders need to understand that a certain degree of inconvenience is anticipated in school renovation projects. Good communication among the designers, the contractor, the school, and the school district central office can keep any possible conflicts to the minimum. It is strongly recommended that some kind of progress chart be developed to coordinate the renovation work at school. The progress chart should include the stages of renovation, the dates that have been committed to, the amount of work to be completed by the contractors, and the actions and reactions of the school district/school. The development of activities on the progress chart has to be agreeable to all parties concerned and copies of the chart need to be distributed to all stakeholders in the renovation process. An example of a progress chart for school renovation is in appendix C.

PRINCIPLES OF SCHOOL RENOVATION

In renovating old school buildings, planners need to follow certain basic principles that relate to the intent of renovation. These principles are practical use, life of the school building, the cost factor, and participatory planning. They are discussed in more detail.

Practical Use

One reason why an old school building needs to be renovated is that the school building cannot adequately accommodate the needs of modern educational programs. For example, the instructional approaches of traditional academic programs, such as math, science, language arts, and social studies, have been changed. Most of them require their spaces to be enlarged or modified to suit their particular activities. In addition, new programs like technological design, multiculturalism, and counseling require new spaces not accounted for in the design of the old building. This means that new spaces have to be created out of old spaces by taking down old walls and erecting new ones. Since the number of square footage is not changed, planners face great challenges to make use of limited funding and resources to achieve the purpose of practical use of the school building.

Life of the School Building

Another reason for renovation is that funds eligible for capital outlay use are not enough for new construction, while the need exists to continue to use existing school buildings. The public expects the school district to invest in old school buildings so that they can be upgraded to serve new educational purposes. Therefore, one of the planners' principles is to ensure that the school building will last for many more years as a result of the renovation work. This means that upgrading the basic structural systems of the school buildings is high on the priority list of things to do in the renovation.

Cost Factor

Because of funding constraints, the long list of proposed work does not usually get finished in a school renovation project. How much money should be budgeted for a school renovation project? The project cannot be justified if it ends up being very expensive. Experts have come up with different ways to weigh if it is worth proceeding with a renovation project. Linn (1952) questioned the worth of a school renovation project if it exceeded 50 percent of the cost of new construction. Castaldi (1994) developed a formula to take into consideration the spread of the renovation costs in the remaining years of the school building after renovation. Other building experts have also examined the conditions of the major building systems in deciding on the value of school renovation. There is not one magic formula that works in all projects to determine if a school deserves renovation. Every school building under renovation consideration has unique attributes that warrant preservation. Therefore, many school districts make the decision to renovate school buildings by using their own professional judgment. The cost is an essential factor for consideration because a substantial percentage of the renovation budget is usually reserved for contingency use.

Participatory Planning

The success of a school renovation project, like any new school construction, depends on the participation of all stakeholders in planning. As mentioned earlier, a planning committee should be formed to include school administrators, facility planners, designers, teachers, school staff, central office staff, parents, and community representatives. Participatory planning is especially important in a renovation project because of limitations in building structure, funding, and resources. The reason for involving all stakeholders is to ensure that the prime interest of each representative group is addressed. And when painful decisions have to be made, the pros and cons of the options can be fully explored by all the participants. Renovation decisions made as a group must be fully understood and endorsed by all the participants.

COMPLIANCE WITH NEW BUILDING CODES

School buildings completed in previous years met the building fire codes of the time. Since then, the codes have been updated and, in most cases, have become more stringent. But school buildings are not required to meet new codes every time the codes are updated until the buildings undergo major renovation. In most school renovation projects, the school district is required to bring the renovated school building up to current codes. The local building department will determine if the extent of the renovation is major and requires a code update. Cosmetic work, like painting and carpet replacement, is not considered a major renovation. By definition, renovation work involving structural changes, such as wall demolition and reconstruction, is considered major renovation work, and is required to meet new codes. These codes include the most updated building codes, fire safety codes, and disability standards.

The local building department exercises authority over the common building codes listed in the standard building code. These codes cover areas of structural, electrical, plumbing, and mechanical work and are developed to ensure building safety. NFPA 101-Life Safety Code, commonly known as the fire safety code, is related to life safety design. This falls within the jurisdiction of the local fire department. The disability codes, known as the ADAAG (1990 Americans with Disabilities Act Architectural Guidelines) and the ANSI (American National Standards Institute Specifications for Making Buildings and Facilities Accessible to and Usable by Physically Handicapped People), are implemented either by the local building department or the fire department. They require handicapped accessibility to all public buildings, including educational facilities. Bringing an old school building up to date to meet current building codes can be very costly. But school districts have no choice but to treat these as priority items in the renovation project. Since all the building codes, fire safety codes, and disability codes are national codes implemented by local officials, the local fire marshal and building inspectors have full authority to interpret and enforce the codes, including exercising discretion over gray areas. Therefore, planners in school renovation projects need to work closely with their local fire marshal and building inspectors in every stage of the renovation to ensure that the renovated buildings meet the current codes and, eventually, their approval.

In addition, school renovation projects need to meet the standards developed by the state department of education. These are a set of regulations established to ensure that schools being constructed or renovated meet the specified educational needs. Regardless of the source of project funding, all design documents for school construction projects need to be submitted to the state department of education for review and approval.

ADVANTAGES AND DISADVANTAGES
OF SCHOOL RENOVATION

In planning for school renovation, planners need to recognize the advantages and disadvantages of renovation projects. Planning practices need to make the best use of available resources to maximize the advantages and minimize the disadvantages. The advantages of school renovation are identified as saving capital outlay expenditures, preserving cultural heritage, and shortening the time in securing needed educational facilities. Renovation projects are disadvantaged in the following areas: hidden conditions, remaining life of school building, and limitations of existing structures.

Saving Capital Outlay Expenditures

A great advantage of school renovation is a big saving on capital outlay expenditures. Most school renovation projects run less than half of the cost of new construction. This means that more funds can be made available to support other school construction projects.

Preserving Cultural Heritage

Many old school buildings have become symbols of harmony and cooperative effort in the community. To small communities, school buildings have been their center of activities. Strong emotional attachment is tied to old school buildings where two or more generations attended school. School renovation projects usually receive positive responses from local communities.

Shortening the Time in Securing Educational Facilities

A school building of traditional construction usually takes from 18 to 24 months to complete, whereas school renovation projects are fast-tracked projects on tight schedules. Renovation activities are closely monitored to create a reborn educational environment in a much shorter time. School renovation receives even greater support in fast-growing school districts because of the quick turn around for project completion.

Hidden Conditions

A great drawback of school renovation is the potential for hidden conditions in an old building. Many structural systems are exposed and are observable, while some are not. Original as-built drawings do not always indicate the real picture of how work was actually performed. Also, unless good maintenance records are kept, evaluating an old school building and its associated systems is difficult. Planners are often surprised by what they find when walls are demolished. Common hidden conditions are the existence of asbestos, rotten support structures, and worn-out wiring and piping that require replacement. Therefore, experienced planners of school renovation projects usually reserve a substantial percentage of the budget to address unanticipated hidden conditions.

Remaining Life of Renovated School Buildings

Objections to school renovation projects usually are focused on the remaining life of the renovated buildings. It is true that the dollars spent on a renovated school need to be justified. Renovation work needs to emphasize improving the structural systems that would prolong the life of the school building.

Limitations of the Existing Structure

An old school building needs to be evaluated to see how it would fit the intended educational programs. The existing design and structure of the building may limit the extent that the building can be converted to new program use. For example, certain load-bearing walls cannot be relocated without substantial cost, and low ceilings and double-loaded corridors may not be redesigned to accommodate open space.

BIDDING AND CONTRACTING FOR SCHOOL RENOVATION PROJECTS

Bidding and contracting for school renovation projects follow the same procedures as a new school construction. The state procurement codes and the school district purchasing guidelines have to be followed. A formal, open bidding process is required for any contract above a set amount, say $25,000, that varies from district to district. The bidding process, bidding conditions, bid preparation, and bid analyses are discussed in the following section.

The Process of Bidding and Contracting

A bid package is prepared to include the bidding requirements and conditions, a full set of drawings for the renovation project, and a full set of technical specifications. Most states will require the bidding announcement be advertised for a minimum of three times in three consecutive weeks. Bids submitted by qualified bidders will be publicly read at the specific time of bid opening. School districts will return all bids that do not satisfy all the required bidding conditions. All satisfactory bids will be fully analyzed to determine the low bidder. Then a recommendation will be made to the school board to award the renovation contract to the low bidder. At the approval of the school board, a notice of award will be issued to the identified contractor, who has 30 days to prepare the contract documents. In addition to signing the formal contract, the contractor needs to submit copies of a valid contractor license, all the subcontractors' licenses, construction permits, proof of workmen's compensation and liability insurance, and a 100 percent performance bond to the school district for review and approval. The school district will then issue the formal notice to proceed to the contractor, who will schedule the start of the renovation work with the school principal.

Bidding Conditions

To submit a formal bid for school renovation projects, a contractor will offer a base bid and alternate prices on top of the base bid with an authorized signature. Together with the bid prices, the bidders are required to turn in a 2 percent bid bond in support of their bid. Some school districts have contractor prequalification procedures that contractors have to successfully go through to qualify for bidding. In the prequalification procedure, the potential bidders' qualifications; building experience; previous collaboration with school personnel, architects, and subcontractors; financial status; and professional references will be examined.

Bid Preparation

In preparing the bid format, planners need to solicit the base bid price, the alternate bid price, and the unit bid price. The base bid price covers everything that is specified to be done in the bid documents. The alternate bid price refers to pricing items of different specifications as substitutes. This is very often put in place in bidding for renovation projects that involve a great deal of unknowns. For example, the base bid may call for vinyl composition floor tiles, and the alternate bid could solicit a price for more expensive ceramic tiles. Alternate bids allow the flexibility of increasing or decreasing the extent of the renovation work from the base bid per budget limitations. Unit bid prices are quoted for work not included in the base bid, but are provided in case work needs to be performed in those areas. For example, any carpet replacement work could cost as much as $25 per square yard, including the cost of new carpet, carpet demolition, and installation. If carpet replacement is not needed in the project, then the unit price for carpet replacement is not needed. But, if it is needed, the price is fixed at $25 per square yard. Alternate pricing and unit pricing are used very frequently in renovation projects because they allow the schools the opportunity to pick and choose the options that best suit their budgets and needs.

Bid Analysis

The base bids, alternate bids, and unit price bids need to be carefully analyzed to see what they really mean to the final price. Decisions have to be made on which alternates need to be accepted according to the budget allowance and practical needs. The acceptance of alternate bid prices could add to or deduct from the amount of the base bid, depending on how the alternates are stated. They could drastically change the status of bidding. Alternate bids are solicited because they provide school districts with the opportunity to make some choices within the budget. Otherwise, if the bids exceed the budgeted amount, time and effort will be wasted in rebidding the renovation project. Unit bid prices also need to be analyzed in conjunction with the base bid price. Some contractors offer a low base bid but high unit bids, but if no additional work is anticipated in those areas, the high unit bids will not be needed. Analysis can be done among bidding contractors by technical areas: structural, mechanical, electrical, plumbing, carpet, tiles, windows, doors, ceiling work, hardware, masonry work, painting, wood work, and equipment. School districts can contact the bidding contractors to clarify their bid prices and supply additional breakdown prices for reference. As a result of the bid analysis, the apparent low bidder will be identified. If the low bidder's price is reasonably within budget, he or she will be recommended to the school board for the award of the contract. However, if the low bidder's price is above the budget, the school board could approve additional funds for the project or reject all the bids and rebid the project. In planning for rebidding, planners will find the analysis of prices from the first bid most helpful. Architects and engineers will examine areas that could possibly be overplanned and revise the package for rebidding.

SUMMARY

This chapter covers the major aspects of work in planning for school renovation projects. It starts with consideration of the extent of school renovation and how priorities are formed. Different scheduling alternatives are then offered for consideration. In planning for school renovation, facility planners follow certain basic principles. Compliance to current building codes is a high priority. The advantages and disadvantages of school renovation are also explored to provide additional information for decision making. Finally, the process of bidding and contracting a school renovation project is fully discussed. The implementation of a renovation project in a school is the topic of the next chapter.

EXERCISES

1. Interview the facility director, or someone responsible for facility planning, in your school district. Discuss the process of developing a budget for school renovation.
2. Do community members really want a renovated school instead of a new school? Design a simple survey to poll your community.
3. Review the last three school renovation projects in your school district to see what exactly was involved in those projects. Any similarities? If not, why not?
4. Use a school renovation project in your school district as an example. Examine what was performed to comply with current building codes. If this work had not been done, what sections of codes would have been violated?
5. How do school districts disqualify contractors with a questionable reputation from bidding a school construction job?

6. Do school districts have to offer a contract to the low-bidding contractor who has a questionable record of performance? If not, what are the legal ramifications?

7. In school renovation projects, planners have to work within the budget amount to convert the building to satisfy many needs. Using the old Batesville High School building in the scenario of this chapter as example, develop a list of items that need to be performed to convert a high school building for middle school use. Prioritize the items on the list. Justify your recommended priorities.

SCHOOL RENOVATION PROJECTS AT WORK

SCENARIO

Recently, school officials in Cloudland District No. 9 decided to remodel Clearwater Elementary School rather than build a new school. Clearwater Elementary was built in 1951 to house 600 students. It was remodeled in 1980 and enlarged to accommodate 750 students. The upcoming remodeling work is to give the building a new look, as well as to increase the student capacity to 1,000. In February 2002, Superintendent Thurman recommended to the board of education that a study be conducted by Building Associates, a construction management company, to determine the feasibility of renovating Clearwater Elementary rather than building a new school. Based upon approval of the board, Building Associates issued a report in June 2002 that outlined the proposed building renovation.

In October 2002, the board decided to issue bonds for the renovation rather than to finance it through capital outlay funds. However, due to financial instability in the country, it was March 2003 before the bonds were issued. Building Associates started the process of finalizing the building plans and a proposed plan was submitted to the board in June 2003. The board granted final approval to the project in July 2003 and asked that construction begin immediately.

Superintendent Thurman and Building Associates met in July to develop a plan for construction while school was still in session. Neither Superintendent Thurman nor any of his staff had experience in building renovation while school was in session. He asked Building Associates to provide recommendations for the renovation and also asked for locations where he could visit other schools that were in a similar situation. But what would he look for? What were the issues that demanded his first priority?

WHAT DIFFERENCE DOES RENOVATION MAKE TO A SCHOOL?

To renovate or to build new is the question that perplexes many school administrators (Castaldi, 1994). The advantages and disadvantages of renovation were discussed in an earlier chapter. However, each district must make decisions about what to do with older buildings (Kennedy, 1999). Is an old school building worthy of renovation? Three questions are obvious: 1) What is best educationally for students? 2) What can be accomplished within the resources available? 3) How much longer would a renovated school building last? The answers to these questions will help provide the school administrative planning team with direction. Renovate or build? Once the decision is made to renovate, what are the important issues that must be addressed before renovation begins (Coley, 1989)? Keeping a school in operation during renovation is quite a challenge to a school principal. The school environment will definitely be affected and routine operational procedures interrupted from time to time. Yet, the school principal has to ensure a safe and healthy environment for student learning.

THE ROLE OF THE SCHOOL PRINCIPAL IN SCHOOL RENOVATION

The operation of school plant facilities includes: 1) maintaining an environment conducive to learning; 2) maintaining the attractiveness of the plant and grounds; and 3) providing maintenance services. The school principal is responsible for seeing that each of these tasks is performed satisfactorily, although some of these services may be administered on a districtwide basis (Kowalski, 2002). A list of "What School Administrators Must Know about Facility Planning and Management" is included in appendix G.

The school plant facilities and grounds must be properly maintained and physically attractive (Bradley & Protheroe, 2003). The principal has accountability and a leadership role with the central staff in seeing that school facilities are properly maintained and functional (Smith & Ruhl-Smith, 2003). The principal must regularly survey the school plant and work with the central office personnel in the maintenance process (Ridler & Shockley, 1989). The administrative staff and faculty must be always alert to any conditions that might be hazardous to the safety of pupils (Healthy Schools Network, Inc., 2000). The principal is functionally responsible, even though many of the services may be under the administration of the central office (Henry, 2000). Any unsafe conditions created as a result of renovation must be addressed immediately. The principal must ensure student safety during the construction or the renovation (Sturgeon, 2000). Safety issues and concerns are particularly predominant during renovation if the school must remain open during the construction phase of the project (Chan, 1996b; DiBella & Anderson, 2000; Jones, 2002). Some researchers have speculated that student achievement drops during periods of renovation. Others have recommended incorporating the renovation into the curriculum and using it as a learning opportunity (Sturgeon, 2000).

In a large, comprehensive high school or middle school, conditions in the building can contribute to injury (Chan, 1996b; Gottwalt, 2003; Sturgeon, 2000). In a senior high school, hazardous conditions in the shops are of special concern (Healthy Schools Network, Inc., 2000). Science laboratories must be free of outdated chemicals and kept in first-rate repair. All chemicals should be kept in safe containers under lock and key (Richardson, Gentry, & Lane, 1994). Regardless of the extent of school renovation, plans have to be made to avoid any hazardous conditions created by renovation.

The states require school administrators to hold fire drills. Because there is always the possibility of fire, fire drills should be held on a regular basis and closely supervised by teachers and administrators (Gottwalt, 2003). The principal should survey the school plant facilities for fire hazards (Richardson, Gentry, & Lane, 1994). A school under renovation has no excuse for not practicing fire drills.

The principal must also be aware of the instructional aspects of the school as they affect the management duties. The principal must examine the following aspects of the instructional program to make sure that they are not negatively affected by school renovation work:

1. The strengths and weaknesses of the instructional staff;
2. Important discoveries and significant advances in instructional programs;
3. Information regarding growth and development of children;
4. Trends and continuing changes in instruction;
5. Problems, needs, and possibilities of instruction;
6. Resource people who can be utilized in instruction; and,
7. The needs of students (Lane & Richardson, 1997).

WORK RELATIONSHIPS: SCHOOL ADMINISTRATORS, ARCHITECTS, AND CONTRACTORS

Consistent expectations, intense communications, and realistic understanding should characterize the working relationships among the architect, contractors, and school administrators (Kendler, 2003). Sound management tools and controls should be utilized and understood by all constituents (Sanoff, 2002). All parties involved need to establish a channel of communication and agree upon a contact network so that not only do they meet on a regular basis to discuss project progress, but they also understand how things are handled in case of emergency (see Communication Chart in appendix I). School principals need to recognize their educational administrative responsibilities, while facility planners, architects, engineers, and contractors handle the technical aspects of school renovation. Many school districts employ project-management firms rather than try to coordinate all the activities of the renovation project (Rochefort & Gosch, 2001). The selection of a project manager should be done with extreme care. Regardless, school administrators must make sure that effective project controls and safeguards are in place and appropriately used (Sanoff, 2002). According to Rochefort and Gosch (2001), these controls should consist of:

a project schedule, agreed upon at the outset and updated monthly; a detailed schedule of values; procurement and materials controls; tracking of quality control and inspection; daily and weekly reports summarizing work performed and problems encountered; recovery plans; and minutes of all project meetings, circulated and approved. (p. 348)

SCHOOL RENOVATION AND COMMUNITY RELATIONS

As with most projects that involve community support, the superintendent should actively involve the community in any renovation effort (Kosar, 2002). The support should be solicited from the discussion phase of the project all the way to the evaluation of the project after completion (Sanoff, 2002). Community involvement is often time-consuming and challenging (Kosar, 2002). However, input and buy-in from numerous and diverse constituencies is critical to the success of any renovation project (Kendler, 2003).

The first question that must be considered is: What is the historical, sentimental, social and political effect of the proposed renovation? Some schools have been in community neighborhoods for 50 years or more and the community considers the school an integral part of the neighborhood (Sanoff, 2002). Often influential citizens are not enthusiastic about the possibility of changing the community school (Stevenson & Terril, 1988).

The planning team must also consider whether or not any outward elements of the school will be changed (Sanoff, 2002). It is often much easier to convince the community of needed modifications if the changes are mostly inside or if the changes do not substantially alter a major part of the school's edifice. Will the trees and other important attributes of the school be maintained? Will entrances and traffic patterns remain the same (Stevenson & Terril, 1988)? In some instances, the community may want to take advantage of the renovation project to incorporate features to accommodate community functions and activities. Early communication with the community will allow an opportunity to have the community's wishes realized.

Community involvement is critical to the renovation project. However, the diversity of the interests also hold potential for delays and budget overruns (Kendler, 2003). The key to community involvement is open communication and early involvement (Kosar, 2002).

MANAGEMENT OF CONSTRUCTION WORK DURING SCHOOL HOURS

If there are no other viable alternatives, then construction should proceed during school hours. However, this must be done with the utmost planning and care (Baule, 1999). Ideally, the renovation would be accomplished without students being present, but the ideal world does not always exist. Consequently, the planning team, before the actual start of construction, should explore several possibilities:

1. Partitioning off the building section by section;
2. Limiting access to the construction site;
3. Completing as much of the construction as possible after school hours;
4. Having the school operate on a four-day week to provide extra time for construction;
5. Saving the most potentially dangerous parts of the project (demolition of parts of the building, etc.) for the close of school;
6. Working with other agencies to secure space for students for one or two days per week (For example, ask a local church to house the first grade on Mondays to permit the construction crew more time in the building and to reduce the amount of occupied space in the building during construction.);
7. Attempting to use a contractor who has experience working in an occupied building;
8. Making sure the community is well prepared for the project;
9. Explaining clearly to parents via the local media any changes in drop-off or pick-up of students; and
10. Inviting the media and the community to view the facility during the construction to keep them up-to-date on the work (some superintendents make videos of the renovation for public relations with the community and board of education).

Again, the key issue during a project of this type is to ensure the safety of the students and the staff. Every precaution should be taken by the planning team and the principal to facilitate the renovation project, while protecting the people in the building (Sturgeon, 2000).

SCHOOL BUILDING SECURITY DURING RENOVATION

School security is a significant issue during school renovation. According to Kendler (2003), some school districts are appointing full-time (24-hour) employees to the construction site. Although this may be expensive at the beginning, the cost savings could be substantial over the term of the project. Some schools employ additional custodial services to maintain 24-hour-per-day presence at the site. Other systems require the contractor to provide security services (Carter, 2003). Regardless of the approach used, the contract should be very specific as to the responsibilities and liabilities of each party, both school system and contractor (Aller, 2002). It is typical that construction projects carry a 100 percent performance bond. Any accident happening during the project construction that results in sustained loss of equipment, materials, time, and completed work will be fully covered. In addition, school districts could work with the local police department to increase patrols around the renovation site to discourage vandalism and break-in attempts.

TECHNOLOGY ACCOMMODATIONS

Most school renovation projects involve some type of accommodation for technology (Butterfield, 1999). The importance of technology in today's education environment cannot be underestimated (Cuban, Kirkpatrick, & Peck, 2001). School administrators must understand the following concepts concerning technology to be able to make viable decisions about the utilization of technology through school renovation:

1. Technology delivery methods (Coley, Cradler, & Engle, 1997).
2. Technology can be tailor-made and customized for special groups (Richardson, Chan, & Lane, 2000).
3. Technology capacity varies by current and future needs (Lane, Richardson, & VanBerkum, 1993).
4. Hardware and software must be compatible (Gooden & Carlson, 1997).
5. Technology education in a political and economic context (Boyles, 1995).
6. Effective utilization of technology takes formative research, segmentation, and augmentation (Cuban, Kirkpatrick, & Peck, 2001).
7. Technology offers a new and challenging method for delivering instruction content (Butterfield, 1999).
8. Technology frees administrators from mundane tasks to engage in the more productive tasks of administration (Richardson, Chan, & Lane, 2000).

The knowledge explosion, particularly knowledge management, is too much for anyone to handle without assistance of technology (Drucker, 1999). Historically, educators have been reluctant to learn and utilize technology, especially computers (Cuban, Kirkpatrick, & Peck, 2001). Teachers are often reluctant to become technologists in addition to being educational professionals (Coley, Cradler, & Engel, 1997). Educators finally agree that the efficient and effective use of technology is becoming a mandate in the workplace (Richardson, Chan, & Lane, 2000).

To many educators, "educational technology" implies devices, not a process approach to instructional delivery (Cuban, Kirkpatrick, & Peck, 2001). Administrators realize they have to break down the barriers and use technology as another instructional tool (Lane, Richardson, & VanBerkum, 1993). To make that happen, technology must be present in the school building for teachers to use (Means & Olson, 1995).

Futurists predict that the education systems of tomorrow will be drastically different from those of today (Drucker, 1999). They forecast that current information about teaching and learning will proliferate and will be used more effectively because of educational technology (Cuban, Kirkpatrick, & Peck, 2001). However, they add that it might take a century to bring to bear the applications of the knowledge base that is now in its genesis stage (Boyles, 1995). After all, they emphasize, it took a century for our nation to develop a superhighway and transportation infrastructure to support the automobile industry that evolved concurrently (Smith & Ruhl-Smith, 2003).

Whether Americans lead the world in this educational transformation or play "catch-up" with more enterprising cultures will in large measure determine whether the United States will continue to have a leadership role in the evermore competitive world of the 21st century (Cuban, Kirkpatrick, & Peck, 2001). It is ironic that a nation that leads the world in technology does not fully employ the same technology in the intellectual development of its young (Richardson, Chan, & Lane, 2000). This phenomenon is of grave concern to business and industry leaders who are unable to acquire an adequately educated pool of employees to fill their workforce needs (Nixon, 1998). They place a strong emphasis on utilizing new technologies to help revitalize schools toward educational productivity (Schofield, 1995).

School leaders need to investigate the cost effectiveness of implementing alternative strategies for delivery of instruction in sparsely populated areas to a limited number of students (Gooden & Carlson, 1997). This includes investigating whether educational technologies provide access to a variety of educational programs that traditionally have not been offered in remote areas. This poses a challenge for educators and their communities to recognize the need for creative and unique approaches to educational renovation and to provide technology services to all students and not just a select few (Cuban, Kirkpatrick, & Peck, 2001).

Technology should be more than computers; it should be all the elements that are utilized by the local school or school district to enhance education efficiency and effectiveness (Richardson, Chan, & Lane, 2000). Planning for technology should be more than planning the number of electrical outlets needed for computers; it should include a projection for future needs (Butterfield, 1999). Will the software match the hardware? Does the school envision a wireless environment? Will students be present in class or in a virtual school at home or some other location? Who makes contributions to the technology? Should technology be about people as well as machines? All these questions and many others need to be answered by the educational administrator planning for technology in a renovated building (Cuban, Kirkpatrick, & Peck, 2001).

PROJECT EVALUATION

One of the items most often forgotten is the project evaluation. Evaluations should be conducted with all stakeholders who participate in the project through all phases of the renovation. These evaluations can be formally structured or semi-structured. The primary purpose of the evaluations is to provide feedback to the superintendent/administrator as the project progresses. If only a formal, end-project evaluation is used, it is too late to make changes in the project. However, if formative evaluations are conducted on a regular basis, then changes can be incorporated as needed or as feasible within resources and facilities (Richardson, Short, & Lane, 1996).

Only those who are active participants in the process or those who have visited the project on more than one occasion should participate in the evaluations. For example, in renovations, the project planning team, administrators responsible for the project, the architect, the project supervisor, etc., should complete evaluation forms. Project evaluation is an important procedure to examine the progress of the project, to resolve potential problems, and to document the renovation status (see project evaluation report in appendix J).

RENOVATION PROJECT ORIENTATION

As a result of renovation, the school building is given a new life. Some of the building's features may have changed: new roof, new floor, new light fixtures, new mechanical systems, new plumbing devices, new boiler, new equipment, etc. Maintenance staff from the school and the central office need to get acquainted with the new systems and their controls. Before substantial completion of the project, orientation meetings should be held between the maintenance staff and the subcontractors of responsible trades to go through the maintenance manuals and answer questions of daily operation. For example, newly installed vinyl composition floor tiles need proper care and maintenance to last; the new boiler temperature checking and daily maintenance procedures have to be followed; the operation of the new dishwasher has to be demonstrated; the functioning of the new fire protection system with the assigned zones has to be well understood; and the new HVAC system needs to be properly maintained on a routine basis to ensure efficient operation. The principal needs to be present at all orientation occasions to learn the maintenance and operation procedures and to establish lines of contact in case of operating difficulties.

WARRANTIES IN SCHOOL RENOVATION PROJECTS

The school district should ensure that the general contractor or the project-management firm has appropriate bonding to complete the project (Stollar, 1967). Contracts between the school district and the general contractor or the project-management firm should be carefully drawn to protect the investment of school funds in the renovation project (Flanigan, Richardson, & Stollar, 1995). Unfortunately, it is at project closeout that most financial problems are actualized (De Haan, 2000). The school district must be sure to have a plan to accept the project without liens and encumbrances. Also, strict guidelines must be in place to guarantee that the renovation is completed on time and within budget (DiBella & Anderson, 2000). Often, performance contracts are drawn-up that penalize contractors and project managers for each day or week that delivery exceeds the due date (Kendler, 2003). Again, the key is to make sure that the contract is very specific and that both parties understand their obligations (Ridler & Shockley, 1989). Most contracts offer a one-year warranty from the general contractors, with possible extensions from manufacturers. Any detected malfunction in the renovation work will be reported to and addressed by the general contractor and the subcontractors during the warranty period. An end of warranty walk-through needs to be scheduled to ensure that all irregularities in the renovation project have been properly handled to the satisfaction of the school district.

SUMMARY

To summarize, the condition, maintenance, and design of school plant facilities influence the people, both adults and students, who are housed in the building. A school building influences the attitudes of teachers, students, and parents. School buildings can have a positive or a demoralizing atmosphere regardless of what the educational leader is attempting to accomplish. The school building is an immediate signal to all about whether or not the administration, the faculty, and perhaps the community, care. Unsanitary conditions cause the spread of disease and illness, depriving students of many days of instruction, and substitute teachers must be called frequently to replace regular teachers who are ill. Conversely, attractive buildings are conducive to the active engagement of teachers and learners and create a joyful atmosphere for all who utilize the building. Renovation will improve the physical conditions of school buildings. However, during renovation, school faculty, staff, administrators, and students have to endure some inconveniences and difficulties that could be resolved if all the renovation stakeholders work respectfully together as a team.

EXERCISES

1. Using computer simulation and CAD, outline a school renovation project with students in attendance.
2. Make a video of a local school renovation project that can be used for presentation to the school board.
3. Develop a technology plan for the school renovation project listed above. Name the data sources. Outline expenditures, revenues, and program concerns.

4. Develop a list of items to be used to judge the "Best Renovated School in the State of _____."
5. Using the brief evaluation instrument included in this chapter, evaluate a renovation project and write a report to the superintendent specifying your findings.
6. Interview at least two school principals who have recently gone through a school renovation experience and report on the strategies they used to ensure a safe and healthy environment for students and faculty during renovation.

SPECIAL ISSUES

9

SPECIAL ISSUES OF SCHOOL MAINTENANCE

SCENARIO 1: HOW TO SAVE ENERGY

Mr. Michael Somersault, the new school superintendent of the Dale County School District, is faced with a balanced budget problem for the coming school year. When he was superintendent of the Rockville County School District, he had launched a districtwide program for energy conservation that saved up to 30 percent of the district's energy consumption. If a similar program could be implemented in the Dale County School District, it would be of great help to him in balancing the budget. Mr. Somersault remembers that the program at the Rockville County School District involved maintaining the operational function of the lighting and mechanical systems of the schools and the central offices. He is not too sure what the Dale County School District has done toward energy conservation. Therefore, he has scheduled a meeting with the director of school maintenance to discuss this matter. What he would like to know is:

1. What is the program implemented by the Dale County School District toward energy conservation?
2. How effective is the current program in conserving energy for the school district?
3. What improvements can be made to the current energy conservation program to make it more efficient and effective?
4. What other new initiatives can be considered in the operation and maintenance of school buildings that contribute to energy conservation?

Mr. Somersault has reviewed literature about energy conservation and summarized success stories of substantial savings as a result of energy conservation to prepare for his meeting with the director of school maintenance. He is anxious to meet with his director of school maintenance and hear what he has to say.

CASE REVIEW

Solutions to energy waste are threefold:

1. Improve maintenance practices;
2. Rework operational routines; and
3. Redesign the features of the building systems.

The redesigning of building systems through school renovation to achieve energy conservation is discussed in chapter 10. This chapter discusses how energy conservation can be achieved by improving maintenance practices and reworking operational routines of schools. As lighting, heating, ventilating, and air-conditioning are the operating systems that consume the largest amount of energy, the following discussion will focus on these particular areas.

Energy Conservation in Lighting

Most school buildings in the United States depend on artificial lights for daily operation. Huge amounts of energy are consumed for a 10-hour operating day. In many schools, classroom lights still continue to burn for a long time after class until custodians have a chance to come and turn them off. Therefore, a substantial amount of energy could be saved if lights are turned off when they are not in use. Several light control ideas are recommended for daily operation of schools:

1. Install motion detectors to turn lights off automatically when no motion is detected in a room;
2. Install a control device to program the lights to be turned on and off at specific times;
3. Educate school children about the good habit of turning lights off when nobody is in the room;

4. Clean light fixtures annually to improve the brightness of lights; and
5. Replace burned out light bulbs with energy-efficient light bulbs.

Energy Conservation in HVAC Systems

Mismanagement in running the heating, ventilation, and air-conditioning systems of a school building can cause substantial energy waste. Energy consumption can be reduced by improving the operation and maintenance procedures of the heating, ventilation, and air-conditioning systems. The following processes are recommended:

1. Replacement of air return filters during maintenance will substantially improve the efficiency of the heating and air-conditioning system;
2. Routine servicing of heating and air-conditioning units by replacing worn-out parts will keep the equipment functioning properly;
3. Program the HVAC system to turn on and off at specific times;
4. Program the heating and air-conditioning unit to run within a specific temperature range;
5. Control the temperature of school buildings during holidays when the buildings are not used;
6. Heat or cool only areas of the building that are used on the weekends or during holidays; and
7. Hot water from the boiler should be maintained between 105 to 140 degrees, depending on the health code of the city or county; a higher temperature is a waste of energy.

Operating the mechanical system of a school building consumes a huge amount of energy. Energy waste can be reduced if a program can be introduced to place more efficient control over the system.

SCENARIO 2: CARPET VERSUS TILE

The carpet or tile controversy has been going on for years. Mr. Sammy Jones, school facility director of the Jackson County School District, is facing a similar dilemma in planning for a new middle school for the school district. Mrs.

McNeal, the newly appointed school principal, is in favor of an all-tile floor in the new school, insisting that her years of experience find that a tile floor is easier to maintain than carpet. Meanwhile, the school superintendent has handed over to Mr. Jones a copy of a research report submitted by a carpet company. The findings of the report indicate that carpet is superior to tile as floor covering for educational use. The school superintendent thinks the findings make sense. Mr. Jones consulted Mr. Queen, the director of buildings and grounds, for his opinion. Mr. Queen said that he did not care as long as the cafeteria was not carpeted. Mr. Jones thought the school custodians may have an answer to whether carpet or tile is a better floor covering for schools. So he visited five schools in the Jackson County School District and interviewed the custodians. Some custodians prefer tile to carpet, and some prefer carpet to tile; the answers have not been consistent. Now Mr. Jones has to make a decision, carpet or tile? The architect is waiting for an answer to complete his drawings and specifications. Mr. Jones knows he has to make a decision soon so as not to delay the design schedule for the project.

CASE REVIEW

Through years of experience in planning and utilization of school facilities, planners know for sure that one kind of floor covering works better than another for certain areas of a school. For example, tile flooring, because of its smooth surface, works better in cafeterias, laboratories, restrooms, kitchens, sink areas, art rooms, storage areas, health rooms, locker rooms, and certain shop areas. Carpet as a floor covering suits some classrooms, theaters, band rooms, choral rooms, and offices because of its acoustical value. The controversial areas are hallways, common areas, and some special-use classrooms.

Basically, designers choose to use carpet because of its elegant appearance and acoustical function. After all, carpet is easy to maintain. School carpet is usually vacuumed daily and shampooed yearly. The problems with carpet are:

- It is not as durable as tile;
- It serves as a bed for dirt and bacteria that is difficult for regular vacuum cleaners to clean;
- It is hard to clean with glues and gums sticking to it;
- Some carpet has a tendency to delaminate when it gets wet;

- Moist carpet is a warm bed for the growth of mildew; and
- Carpet repair work often turns out to be unsatisfactory.

Tiles are chosen as floor covering because certain areas in a school building need to be cleaned up more often than others, particularly wet areas. Tiles are also easy to maintain. Tile floor needs to be mopped daily and stripped and waxed once or twice a year. Lobby areas with tile may need more work to keep them shining everyday. The factors that work against the use of tile as floor covering are:

- Walking or standing on a tile floor is not as comfortable as on carpet floor;
- Tiles have low acoustical value; and
- Wet tiles pose potential danger because of their slippery surface.

When carpet is dry, it lasts and is less likely to generate bacteria. Chan, Richardson, and Jording (2001) offer tips to keep carpet dry:

- Inspect roofs, walls, windows, and doors of the school building for leaks;
- Inspect the crawl space, if any, for excessive dampness;
- Inspect the outside grading along the building perimeter wall for positive drainage from the building;
- Install a vapor barrier when constructing concrete floor slabs;
- Do not excessively wet the carpet during shampooing and hot water extraction;
- Keep the air-conditioning on at night;
- Control the air humidity in the classroom;
- Clean/dry up any spillage immediately;
- Inspect any damage to steam lines and water lines in the building;
- Do not install carpet close to sinks or water fountains;
- Do not install carpet in the cafeteria; and
- Install carpet with a synthetic unitary back.

When the decision is made to install tile as floor covering, then the choices are vinyl composition tile, terrazzo tile, ceramic tile, and quarry tile. Vinyl composition tile is the most commonly used because it is more economical. It does not last as long as the other tiles. Terrazzo tile is the best quality tile on the market in terms of aesthetics and durability, but terrazzo is the most expensive. Quarry tile has a hard and abrasive surface most suitable for kitchen, bathroom, or utility areas. Ceramic tile offers both toughness and attractiveness as floor covering. Some ceramic tiles have an abrasive surface for commercial use. The selection of tiles is based on the needs of certain areas. Aesthetics, budget, durability, and maintenance are the considerations.

SCENARIO 3: DOES STANDARDIZATION MAKE SENSE?

Dr. Thomas, assistant superintendent of Long County Schools, has been to a professional conference about school business management. He attended a workshop on standardization as applied to school district operation. He is so impressed with the idea he believes that if Long County Schools can standardize the way it operates, it would save a lot of time, trouble, and millions of dollars in the long run. Standardization can be applied to procedures, equipment, supplies, and all types of building design. This comes at the right time because of budget problem in the Long County Schools. The public has been complaining about waste in the operation of the school district. Dr. Thomas wants to start with his division as an example to demonstrate the possibilities of applying the concept of standardization. He is particularly interested in stocked items in the school district warehouse and the use of materials, equipment, and design features in the daily operation of the schools. He has called upon his directors of maintenance, facility planning, and purchasing to examine what they could do in their areas to implement the concept of standardization to improve the efficiency and effectiveness of their business operations. He further emphasizes that recommendations to him need to include full justifications of their evaluation of department needs and the benefit to the school district as a public entity.

CASE REVIEW

The idea of standardization originates from the thinking about saving time and effort in the handling of a task. It is very often employed in the business field to improve work efficiency and effectiveness through a uniform process. For example, a business corporation may choose to use a certain platform of computers throughout their offices as a means of standardization. Other examples are the use of standard-size containers for shipping in ocean liners, standardized forms for visa applications, and standardized sizes and shapes of mailboxes. The advantages of standardization are:

- It facilitates the operational process by not having to adjust to different formats every time the process is performed.
- It shortens the time spent on determining how different formats are to be handled.
- Standardization makes it easy for evaluation of products on a comparative basis.
- Standardization facilitates maintenance efforts.
- Standardization makes replacement of parts easy.
- Standardization results in increased orders of single items. The unit price per item of purchase will come down as a result of economies of scale.

Standardization works in many businesses and industries simply because it promises results. But, under certain circumstances, it may not work as well. Standardization has its disadvantages:

- With standardization, only one product or process is specified, which removes the opportunity of competitive offering.
- In some states, proprietary specifications may be challenged in court.
- In many fields, specification of standardization simply promotes monotony and kills creativity.
- Standardization eliminates choices.
- Standardization is opposed to individualization and diversity principles being advocated in education today.
- Standardization may mean big savings for large organizations. But the margin of savings may be minimal for small organizations.

Standardization can be implemented in school facility maintenance and operation in many ways. First, standardization in school maintenance equipment and supplies is definitely a large money saver. Second, standardization in school HVAC systems helps eliminate many maintenance nightmares. Third, standardization in carpet and floor tiles in schools facilitates the storage of replacement materials. Fourth, standardization in the work procedures of school custodians meets fairness standards. Fifth, standardization in the appropriation of resources for custodial services establishes guidelines for meeting the needs of the schools.

Other standardization applications can be seen in the use of ceiling tiles, light fixtures, locks, and paint.

Standardization benefits a school district maintenance department in the stocking of needed materials, supplies, and maintenance equipment. School district maintenance staff can be well prepared to perform similar work from school to school. Lower bid prices for maintenance materials, supplies, and equipment are anticipated with the purchase of larger quantities and waste is kept to a minimum. Standardization is an answer that the maintenance department can provide to achieve higher efficiency in school operation. However, barriers stand in the way to standardization under certain circumstances, including:

- The maintenance department is always at the receiving end of the facility planning process, although its input to planning is solicited at the beginning. At the end of the warranty period, a new school building with many nonstandardized features may be handed over to the school district for maintenance.
- School communities may push for individualization of school buildings in their community while shunning standardization.
- Donated materials and equipment do not usually meet the school district requirements for standardization.
- Creation of new products and discontinuation of obsolete items by manufacturers make standardization difficult.
- It is not easy to maintain standardization in school districts with school buildings of varying ages.

To sum up, standardization is certainly an approach to save both time and effort in school maintenance. However, not all areas of school maintenance can be standardized in a manner that is beneficial to the school district. In many instances, the standardization of maintenance work is not even under the control of the maintenance department. In general, the department has better management of standardization in specifying maintenance supplies and equipment.

SCENARIO 4: DEALING WITH VANDALISM

Mr. Green is the new principal of Hoover Elementary School, which is located in a community consisting mostly of low-income minority families. Eighty percent of his students are minorities and more than half of them are eligible for free school lunches. Mr. Green is faced with several challenges at school: below average student achievement, poor parent participation, and many behavioral problems. Above all, what bothers him the most is the increasing number of acts of vandalism to school property, both inside and outside of the school building. One of Mr. Green's priorities to bring improvement to the school is to maintain a positive image of the school building by fighting vandalism. He does not believe that vandalism has anything to do with minorities and low socioeconomic status. He trusts that his determination to reduce vandalism could make the difference. Mr. Green starts his effort by establishing an anti-vandalism committee at school, consisting of teacher representatives, the school's head custodian, the school district maintenance director, parents, and his assistant principal. His committee is committed to fighting vandalism at the school.

CASE REVIEW

School properties are being vandalized every day, costing school districts substantial amounts of their budgeted dollars to continue to repair damages. Budgeting for repair to vandalized school buildings is difficult because of the unpredicted elements involved, such as frequency and severity. In addition, addressing damages to school property as a result of vandalism takes up a tremendous amount of time of the school custodians and school district maintenance staff.

Cooze (1995) identified the three possible reasons for school vandalism:

1. Those associated with criminal acts (e.g., breaking windows or doors during a burglary);
2. Those constituting acts of retribution (e.g., an angry student retaliating against the school because of some grievance);
3. Those constituting senseless destructive behavior (e.g., individuals who simply obtain gratification from destroying another person's property).

In some cases, school property is vandalized as a result of a student confrontation. Older students with greater strength can cause more severe damage than younger students. As a result, there are more vandalism occurrences in high schools and middle schools than in elementary schools.

Kowalski (2002) associates school property exposure to vandalism with several variables, such as district economic conditions, district location, district size, building condition, security provisions, and maintenance programs. Though more research is needed to provide evidence to support these associations, the connection of vandalism to these variables seems apparent.

The interior of a school building is most likely to be vandalized during the school day. Breaks, lunch hours, class changes, and after-school hours are the most likely times when vandalism occurs. Vandalism to school property can also happen during class time when teacher supervision is lacking. Nights, weekends, and holidays are the times when the exterior of a school building is most often vandalized by burglars, trespassers, and unwelcome visitors.

The areas inside a school building most vulnerable to vandalism are restrooms, recessed corners, remote areas, science laboratories, hallways, and locker rooms. Restrooms are the number one target for vandalism. Vandalization of tissue paper holders, soap dispensers, paper towel holders, plumbing fixtures, toilet partitions, water faucets, mirrors, light fixtures, ceiling tiles, and even smoke and heat detectors inside the restrooms are frequent occurrences. Other items in the school building subject to vandalism include ceiling tiles, floors, locks, doors, walls, furniture, equipment, lights, and fixtures.

School property outside the school building, especially hidden areas at the back of the building, are also subject to damage by vandalism. Breaking windows, setting doors on fire, painting graffiti on walls, breaking in, ruining rooftop equipment, and defacing school signs and exterior lights are some of the most common vandalism activities. Additionally, portable classrooms or separate storage buildings on campus are often the targets of vandalism because they are not protected by intrusion alarms and their locking devices are usually very flimsy.

Vandalism cannot be totally prevented. Every school building will be vandalized to a certain extent every year. However, conscientious effort led by school administrators can substantially reduce the number of occurrences. The school principal in the scenario at the beginning of the chapter is heading in the right direction by establishing an anti-vandalism committee charged with the responsibility of fighting vandalism. Some of effective measures could include:

- Warning approach: Install "No Trespassing" signs and warning signs of a burglar alarm in action.
- Detecting devices: Install a burglar alarm, security lights, and a surveillance system inside and outside of school buildings.

- Vandal-resistant designs: Design school buildings with vandalism prevention in mind. Specify durable materials for school construction.
- Administrative measures: Establish policies for managing cases of vandalism. Specify students' rights and responsibilities in using school facilities.
- Community support: Organize a community watch effort to keep an eye on school property. Request that the police department increase patrols around school property at night and over the weekends.
- Repair work: Work with school custodians and the school maintenance department to arrange repair work soon after damages occur.
- Campaign effort: Promote a sense of belonging in caring for school property through a campaign effort.
- Supervisory capacity: Increase the teachers' supervisory capacity over student activities by extending supervisory time and enlarging supervised areas.

SUMMARY

Four major school maintenance issues have been discussed in this chapter: energy conservation, carpet versus tile, standardization, and vandalism. The scenarios provide situations faced by many school district administrators. In suggesting hints to handle the problems in the scenarios, the authors have attempted to discuss the issues from both the idealistic and realistic perspectives. While this chapter cannot discuss the many maintenance issues that could occur, evaluation of the ideal and the reality will clearly indicate a practical solution to these unique problems. The questions in the following exercise are intended to provide readers with additional challenges.

EXERCISES

1. Suppose that your school is planning for replacement of floor covering the next summer. As assistant principal chairing the planning committee, evaluate the need for carpet or tile in all areas of the school. Make a recommendation based on budgetary, durability, and aesthetic considerations.

2. Which is an easier floor covering to maintain, carpet or tile? Analyze the issue by reviewing what literature has said and also by interviewing at least five school custodians to derive a conclusion.

3. Do you agree that standardization makes sense in school maintenance? Why? Give specific examples to justify your argument.

4. Identify all the areas of school maintenance responsibilities and examine how the standardization concept can be successfully applied to each of these areas.

5. In collaboration with the maintenance department in the central office, develop an energy conservation plan for your school to involve faculty, staff, parents, and students.

6. Do energy conservation strategies work in your school district? Conduct a study in your school district to document the effectiveness of these strategies.

7. Refer to the case in Scenario 4. Follow the effort of Mr. Green in fighting school vandalism. What would you do next if you were in Mr. Green's place? Draft an anti-vandalism proposal to reduce the number of vandalism activities in his school. Conduct a study to show the annual cost of building repairs caused by vandalism in your school district.

10

SPECIAL ISSUES OF
SCHOOL RENOVATION

SCENARIO 1: LONG-RANGE FACILITY PLANNING

Jenkins Elementary School and Williams Elementary School are 35- and 31-years-old, respectively. Both schools, located in the Evans Community of Larks School District, have been scheduled for major renovation for the next two to three years. However, a recent study of demographic trends showed a tremendous increase in the middle school population in the last few years because of immigration. The trend is projected to continue in the next 5 to 10 years. Because of this unanticipated growth, the long-range facility plan of the Larks School District has to be revised. The proposal for revision includes postponing the renovation of Jenkins Elementary School and Williams Elementary School. The fund reserved for the Jenkins and Williams school projects is proposed for transfer to the construction of a new middle school to meet the anticipated needs of the community. Since the new middle school is also located in the Evans Community, some school board members think that the community will not be shortchanged because the promised capital outlay fund will remain in the community but will be used for the new middle school project. Mr. Ralph Bennett, the school superintendent, also shared similar ideas that the proposal will not meet with much opposition. But the school board and the school superintendent underestimated the situation. When the parents of Jenkins Elementary School and Williams Elementary School heard of the proposal, they were upset that the school district failed to fulfill its commitment to elementary education in the Evans Community. They are planning to boycott the middle school project. Some of them are also threatening a legal challenge to the school district.

CASE REVIEW

Long-range facility planning for a school district starts with student enrollment forecasting, which is performed annually to review the growth patterns of the student population in the school district. Student enrollment forecasting is usually done for the next five years and then the 10th year. Enrollment forecasting usually includes districtwide forecasting and forecasting by school and by grade. The feeding pattern of schools is taken into consideration in the forecasting process. The enrollment forecasting will result in a growth rate, a decline rate, or a stable condition of student enrollment. At the same time, a districtwide facility inventory needs to be taken to show the capacities and the physical conditions of the school buildings. The next step is to merge the student enrollment forecasting data together with the facility inventory data. The result will demonstrate whether existing schools can accommodate the forecasted enrollment (see appendix D).

In planning for student accommodation, the age and the conditions of school buildings have to be considered. In growing school districts, all available data need to be converted to indicate the quantity and the location of classroom shortages. Then the decision has to be made to handle the classroom shortage problem by building new schools, by adding new classrooms to existing buildings, or by renovating existing old buildings. Most school districts manage their immediate needs by hauling in trailers before permanent classrooms are built. Then a priority list of capital outlay projects is drawn up to indicate the time sequence of all the projects, including new construction, additions, and renovations (see appendix E). Decisions on priorities are based on the student population growth and the availability of facilities in each area of the school district. The priority list is subject to revision every year by using the most recently generated enrollment forecasting data.

What about school districts that did not experience growth or are projected to not have any growth in the future? Student enrollment in these districts are either stable or declining. The facility needs of these school districts are justified by building new schools to replace the old ones, by initiating new educational programs, by consolidating schools, and by extending the life expectancy of existing school buildings. In school districts of stable or declining student enrollment, the priority of capital outlay projects will also show newly designed replacement schools, additions to existing schools because of program expansion, and major renovations to existing school buildings.

An old school building has to be fully evaluated for renovation before it is placed on the priority list of capital outlay projects. Some school buildings never make it to the priority list despite strong emotional attachment to the building. Where a school renovation project stands on the priority list depends on several factors:

- *Physical conditions of the school buildings.* Buildings with worse conditions are a higher priority.
- *Urgency of needs.* Buildings with more urgent needs are a higher priority.
- *Growth rate.* Buildings located in areas of faster growth are a higher priority.
- *Collaboration with high-priority projects.* Buildings collaborated in planning with other high-priority projects are a higher priority.
- *Funding availability.* Buildings with additional sources of funding besides the state are a higher priority.
- *Time of commitment.* Buildings with longer years of previous school board commitment are a higher priority.

The priority list of capital outlay projects is a sensitive list. It represents a commitment of the school board to the general public. It serves as a guideline for implementation of the capital outlay program of the school district. The list is also the result of many tedious hours of compromises among opposing parties. Every year, the priority list is reviewed by the facility planning committee for possible update.

SCENARIO 2: RENOVATING A PLAYGROUND FOR CHILDREN'S SAFETY

The Sullivan County School District is aware of the issue of playground safety nationwide and the federal regulations on safety codes for children's playgrounds.

To comply with the federal regulations, the superintendent planned to install "fall zones" for student safety in the 34 elementary schools of Sullivan County. Because of budget constraints, he recommended that the playground safety project be divided into two phases. Phase I Playground Safety Project included 20 elementary schools in this year's budget, and Phase II Playground Safety Project included the remaining 14 elementary schools in next year's budget. The Phase I project was completed by the middle of the year. However, the superintendent received two reports of student playground accidents soon after:

1. A student fell from playground equipment in a school where the safety zone had been installed. He was badly scratched by the tree branches mixed with the pine bark.
2. A student was injured when he fell from the playground equipment in a school where a safety zone had not been installed.

The superintendent was upset. He called the business manager for advice so that he could be prepared to meet the angry parents. As the business manager, what would you suggest to the superintendent?

CASE REVIEW

Playgrounds have been constructed for the fun and physical development of school children for years. Not much attention was given to playground safety until records showed that approximately 170,000 children were injured on school playgrounds each year (Thompson, 1991). Guidelines were developed to address the safety of outdoor athletic facilities, including playgrounds and equipment (U.S. Department of Education, National Center for Education Statistics, & National Forum on Education Statistics, 2003). Safety standards were tightened to cover minimum requirements for manufacturing play equipment and designing and constructing children's playgrounds (U.S. Consumer Product Safety Commission, 1981, 1993). School districts were then under public pressure to renovate school playgrounds to meet the most current safety codes. Like many legislative mandates, compliance with playground safety codes is one of those nonfunded issues imposed on local school districts. The situation illustrated in Scenario 2 actually describes the dilemma and difficulties many school districts encounter when they are trying to comply with the playground safety codes. As a matter of fact, most of the playground safety problems are centered on design, construction, maintenance, and supervision.

Design

If a children's playground is carefully designed, it will help tremendously in reducing the number of possible accidents. Some of the design considerations are: playground location, playground size, age group to be served, goals to be achieved, and equipment to be selected and located.

Construction

The quality of work in construction is essential to the creation of a safe playground for children. Special care has to be taken in installing equipment per manufacturers' specifications.

Maintenance

A well-designed and quality-built playground has to be effectively maintained to be functional. The conditions of a playground will run down quickly if it is not well maintained. Maintenance work should focus on keeping a high standard of cleanliness and safety.

Supervision

Many accidents on playgrounds are related to lack of supervision when children are around. A danger warning sign does not serve the supervision purpose. Children on a playground must be closely supervised by either teachers, paraprofessionals, or designated volunteers.

To maintain a high level of safety on playgrounds, Chan (1988b) originated an effective program for playground safety at school. The program calls for a collaborative effort to involve teachers, custodians, parents, and administrators in an accountability model. Any sign of suspected malfunction has to be reported to the district maintenance department for inspection and repair.

With reference to Scenario 2, the school district has tried its best to respond to the playground safety mandates. Unfortunately, the school district's financial status did not allow the completion of safety zones in all the schools in the first year. Obviously, a lack of quality control in the materials used in the fall zone complicated the problem. A lesson learned from this scenario is that educators cannot afford to take a chance in allowing unsafe playground facilities to continue to serve children under their care.

SCENARIO 3: UNKNOWN CONDITIONS IN RENOVATION

Columbia Elementary School, a 40-year-old facility, was scheduled for renovation in the summer months when the students were out of school. The school was expanded 18 years ago and minor renovation work was performed on the original building at the same time. The work in this second renovation includes wall painting, floor tile replacement, carpet replacement, and new ceilings for all classrooms and hallways. Two classrooms will be remodeled as a special education suite. In addition, all restroom facilities in the school need to be renovated to meet the code for accommodating disabilities. In evaluating the nature of the renovation work, the director of facility planning estimated that the renovation project should be completed in 9 to 10 weeks in the summer and that the school should be ready to be opened in mid-August. When the spring semester came to an end in May, the architect and the general contractor were mobilized to start the project. Materials needed for renovation were ordered and scheduled for delivery. Things were running smoothly until the general contractor started demolition work. First, in removing old restroom fixtures, the plumber found that the old waterlines had rusted and corroded so much that they would not meet the current plumbing code. The engineers concurred and recommended that all rusty waterlines in the school be replaced. The plumbing subcontractor asked for $10,000 to do this additional work and seven days time extension. Second, in taking down the wall between the two classrooms to be converted to special education use, asbestos was detected in the drywall panel. Asbestos specialists were called in for further inspection and abatement procedures were strictly followed. As a result, the east wing of the school was closed for two weeks for asbestos abatement. The effect on the renovation project was that it was three weeks behind schedule and $45,000 over cost. The parents were upset that the school could not be reopened on time. The school superintendent was not happy with the way the Columbia Elementary School renovation project was handled and asked to meet with the director of facility planning to come up with a better solution.

CASE REVIEW

Hidden conditions are the worst enemies of a school renovation project. They come as a surprise and end up delaying the project and incurring cost overruns for the district. When the decision is made to renovate a school facility, surveys and inspections of all types are conducted to collect data on the existing conditions of the school building. Original construction drawings and maintenance records are carefully reviewed to determine the best approach to connect the proposed renovation work with the existing building systems. Based upon the as-built drawings, estimates will be made of what could possibly be buried underground, embedded in the walls, and hidden above the ceiling. However, in reality, schools may not be built exactly like the drawings. This is because of the following reasons:

1. Some drawings are unclear. Contractors have to refer to the architects and engineers for clarification or use their best judgment to build the school.
2. Some drawings of renovation projects do not reflect the exact conditions under which the contractors had to work. Contractors will adjust their work to the existing conditions through their experience.
3. Conflicting directions between the drawings and specifications, or between sections of the drawings and specifications, cause confusion. Clarification usually means a change from the plan, and change addenda are not always appended.
4. Contractors sometimes choose to install a system in an alternate way because of convenience or cost saving.

In all of the above conditions, the project manager should record the changes in the as-built drawing to be handed over to the school district at the end of the project. But, very often these changes are not properly recorded or recorded at all. These are the dilemmas of working with an existing school building, particularly an older building in which work has been done over the years.

Facility planners of school renovation projects sometimes have to plan from what information is available to them and deal with the situation when unknown conditions are revealed. Strategies facility planners use to manage unknown conditions of school building renovations include:

- Do a thorough inspection of the school building and carefully evaluate the available data for the facility;

- Leave a large proportion of the proposed budget for contingency use; 20 percent of the budget is appropriate;
- Allow more time for renovation; it is risky to run a tight schedule for project completion;
- Explore options in managing crisis by consulting the architects, the engineers, and the director of school maintenance;
- Delaying the school's opening is not good publicity; try to work around this community-sensitive issue by looking at what it takes to keep the project on schedule; and
- Confirm assistance from the director of school maintenance and the school district purchasing agent. They are the key people to turn things around in a time of emergency.

In planning for school renovation, school administrators should anticipate that unknown conditions will be revealed. The key element is to act fast, determine the most logical options to pursue, secure all necessary resources to make things happen, and keep the renovation project going.

SCENARIO 4: SCHOOL RENOVATION FOR ENERGY CONSERVATION

Mrs. Susan Owens, superintendent of the Ryan County School District, returned from a regional superintendents' meeting a week ago. She is surprised to know that the Turner County School District, a rural district similar in size to the Ryan County School District, consumed only half of the energy of the Ryan County School District. She is bothered by the apparent waste in her school district. In her previous experience as school principal in another school district, she had gone through a successful school renovation project that addressed the energy conservation issue. The Ryan County School District maintains one high school, two middle schools, and six elementary schools. She is interested in pursuing a districtwide energy conservation program that would save the school district a considerable amount of money in the long run. Mrs. Owens has reviewed some significant literature on energy conservation for schools and has visited three school districts in the state to examine their energy conservation projects. Determining that this is the pressing issue of the coming year, she called upon her director of buildings and grounds and all the school principals to meet to discuss planning for energy conservation in the school district.

CASE REVIEW

The U.S. Department of Energy estimates that out of the $6 billion spent annually by school districts on energy consumption, $1.5 billion could be saved (Reicher, 2000). Energy consumption comes to the attention of school administrators because it can save a large amount of money in a time of budget difficulties. A districtwide energy conservation program can be very extensive. It starts with an energy audit of all the school facilities in the school district (Castaldi, 1994). Energy consumption by type and by building is examined. Consideration is given to the kinds of equipment and models used. Energy consumption data are compared between schools based on building square footage or the number of occupants. Comparison is often done between school districts of comparable size within the same geographic zone. Kowalski (2002) recommends that energy audits include data from three cumulative years to come up with an accurate assessment. Through comparative analyses of data, energy audits identify schools with excessive energy consumption.

The two largest sources of energy consumption in schools are lighting and the operation of the mechanical system that includes heating, ventilation, and air-conditioning. Other energy consumption involves operating AV equipment, computers, shop equipment, the surveillance system, the burglar alarm system, the P.A. system, maintenance equipment, and office equipment. Most energy conservation programs focus on lighting and mechanical system operation.

According to Kowalski (2002), a school district energy conservation program incorporates three categories:

1. Solutions requiring alterations in operations;
2. Solutions requiring alterations in maintenance; and
3. Solutions requiring design changes.

Solutions to energy waste problems requiring design changes are addressed in this chapter. Other solutions to energy waste were addressed in chapter 9.

Energy Conservation in Lighting Design

Lights in schools consume a large amount of electricity. During the energy audit, records should be kept of the foot-candles measured in different areas of a school. Some light fixtures should be removed from rooms where excessive foot-candles of light are measured. Some of the latest models of fluorescent light fixtures are manufactured with energy efficient features. The reflective interior of the fixtures tremendously improves the light intensity. Replacement of light fixtures to achieve a savings has to be carefully considered in terms of payback time. Other lighting design features were suggested by MacKenzie (1989), including multiple light fixtures in rooms with windows so that lights by the windows are controlled by separate switches and equipping multiple entrances to rooms with their own switches. Motion-detector lights and timers should also be installed.

Energy Conservation in Mechanical Design

When heating, ventilation, and air-conditioning units run down, they need to be replaced with more energy efficient units. Boilers, whether electric or gas, consume a tremendous amount of energy. Since a new and more efficient boiler is expensive, the replacement cost has to be compared and analyzed with the operating cost. In most places, gas is cheaper than electricity. Therefore, in replacing equipment, preferential consideration needs to be given to gas-heated equipment. Additionally, the exhaust hood in the school kitchen pulls a tremendous amount of air from the inside to the outside of the building. A mechanical device to make up at least 90 percent of the exhausted air is needed to maintain school building temperature.

Energy Conservation in the Building Shell

The school building itself has to be energy efficient to maintain the intended interior temperature. Energy conscientious renovation work could include the following items:

• Replace old windows with better insulated double-pane windows.
• Reduce the window space by installing insulated panels.
• Lower the classroom ceilings to reduce the volume of classroom space.

- Weather-strip all exterior doors.
- Insulate all ceiling areas.

Through renovation, a school building can become more energy efficient and save money for the school district in the long run. Energy conservation through school building maintenance and operation was discussed in chapter 9.

SUMMARY

This chapter is devoted to the discussion of four special topics in school renovation: long-range facility planning, playground safety, unknown conditions, and energy conservation. School districts are encouraged to prolong the life of their existing facilities through renovation. They need to justify the worth of renovation work and set priorities. During renovation, safety conditions and energy conservation concerns need to be addressed. However, unknown conditions exist in every school building. They must be properly addressed as they are uncovered. Special attention is called to playground safety, a federal mandate to be complied with by all school districts.

EXERCISES

1. In Scenario 1, when a capital outlay fund was previously approved by the state for the renovation of the two elementary schools, does the school district have the authority to redirect the fund to other projects? Why? How does it work in your state?
2. In Scenario 1, how do you react as one of the principals of the two elementary schools funded for renovation? How could committed funds be used for other construction projects? What can you do to protect the interests of your school?
3. In Scenario 2, because of budget constraints, the school board, upon recommendation of the school superintendent, approved the motion to delay installation of half the playground safety zones until the following year. In your opinion, is the school board running a risk? If you were the

school superintendent, would you have made the same recommendation to the school board? What other alternatives do you recommend?

4. If you were the school superintendent in Scenario 2, what would you do to handle this situation when it is clear that the school district is negligent?

5. Is there a better way to manage unknown conditions in renovation projects? What would you do differently if you had been assigned to administer the Columbia Elementary School renovation project in Scenario 3?

6. Study a few energy retrofit projects in your school district. Examine the energy efficiency of the renovated school buildings. Are the initial investments paid back in the projected number of years?

7. Analyze the energy consumption of your school district by school building. Can you tell which school building is the least energy efficient? What would you recommend to do in that building to conserve energy?

11

FUNDING FOR RENOVATION AND MAINTENANCE PROJECTS

SCENARIO

The Durham County School District is located in a suburb of a medium-size city in the South. It has a student population of 18,000. The growth of the county was stable until recent years, when some regional businesses moved in and erected plants. The chamber of commerce is monitoring information about 10 other industries expressing an interest in investment in Durham County. The school district planning director has projected an average student population growth of approximately 2 percent for the next 10 years. With this unprecedented growth, Mrs. Waller, superintendent of the Durham County School District, is mobilizing her cabinet to form a Strategic Planning Committee to assess the possible effect of this growth on education and to identify sources of revenue to address educational needs. What Mrs. Waller worries about most is the shortage of educational facilities to house more students. More than half of the schools in Durham County are more than 35 years old. They are close to either complete rundown or need a major renovation. In addition to the need for school replacement, two new elementary schools and one middle school will be needed in the next five years to meet the population growth. The planning director has estimated a total of $50 million in school facility needs in the next five years. The finance director has cautioned that with the current formula for state capital outlay funding, there is no way that the Durham County School District could catch up with the growth without alternative channels of funding. Of the available alternative funding sources, Mrs. Waller is in favor of an additional sales tax, which she thinks will be drawn mainly from shoppers in the neighboring counties. Other sources of alternative funding for school facility construction under consideration include working with the building authority, the building contractor, and private donations.

COSTS OF SCHOOL CONSTRUCTION, RENOVATION, AND MAINTENANCE

The cost of constructing new educational facilities is enormous and puts great pressure on local school systems to find funds (Stollar, 1967). The cost of construction goes up exponentially with fluctuations in the local, state, and national economies (Wood, 1986). School administrators must critically examine the costs of construction for new buildings as compared to renovation and maintenance before beginning any capital outlay project.

The cost of school construction in recent years was summarized by Argon (2000) in the following:

Elementary School: constructed for 610 pupils, 115 square feet per student at a cost of $118.81 per square foot or $9.2 million. Average size is 65,379 square feet and contained 30 classrooms.

Middle School: constructed for 800 students, 130 square feet per student at a cost of $126.86 per square foot or $14.3 million. Average size is 119,000 square feet and contained 36 classrooms.

High School: constructed for 804 students, 153 square feet per student at a cost of $139.48 per square foot or $21.4 million. Average size is 35,559 square feet and contained 35 classrooms (p. 30–45).

As the cost of construction escalates, school administrators need to consider the best investment of tax dollars in school facilities. Depending on the scope of the renovation project, the cost-per-square-foot of renovating a school building usually runs much less than half of the cost-per-square-foot of constructing a new school. If the cost of a school renovation project goes higher than 50 percent of the cost of new construction, the renovation project will face serious problems in getting the school board's approval without substantial justification. Therefore, in terms of stretching the school construction tax dollars, school renovation makes sense. In 1999, public schools spent $5.07 billion on additions and $5 billion on modernization and upgrades (Argon, 2000). In addition, renovation decisions are also tied in with maintenance costs of school buildings. One of the purposes of renovation is to improve the maintenance efficiency of the school building. However, consideration has to be made of the amount of investment in school improvement. Some projects may not be paid back during the remaining life span of the school building (Chan, 1980).

FUNDING FOR SCHOOL RENOVATION AND MAINTENANCE

Typically, the funding for renovation and maintenance projects is accomplished through one or more of the following options (Thompson, Wood, & Honeyman, 1994):

- State funding options
- Local funding options
- Alternative strategies

Historically much of the responsibility for funding school facilities was vested in the local district (U.S. Department of Education, 1999; Wood, 1986). Federal funding has not been much involved in school building except for special projects for vocational or disability interests. However, as states have assumed more responsibility for education, many states have taken a more active role in building school facilities (Thompson, Wood, & Honeyman, 1994; U.S. Department of Education, 2000).

STATE CAPITAL OUTLAY PROGRAMS

Many states are now responding to challenges to the educational funding provided by the state (Geiger, 2001). The classic example is Kentucky, which created a new educational system based upon the state supreme court ruling that the state system of funding was unconstitutional and did not meet the "fair and appropriate" requirements of the state constitution. Texas and Kansas are two other states that have faced challenges to the financing of education in the state (Thompson & Wood, 1998). Tennessee is currently embroiled in a court case brought by the smaller school systems that believe they are not funded at the same level as the larger school districts (Ritter & Lucas, 2003). The next round of challenges will most likely focus on school facilities. Because facilities have historically been funded at the local level, many states are concerned that the need for school facilities will drive the next generation of school funding legal challenges (Hack, Candoli, & Ray, 1998; Hansen, 1992; Leibowitz, 2001; Rittner-Heir, 2003).

Education has been and is the responsibility of the state (Alexander & Salmon, 1995). According to Alexander and Salmon, "a child is a child of the state;" meaning that every child deserves the same treatment and opportunity as every other child in the state (1995, 146). Voter resistance to approving local bond issues has forced many states to become more active in entering the bond market to provide assets to local school districts (Salmon & Thomas, 1981; Swanson & King, 1997).

Bonded Indebtedness

Bonded indebtedness at its simplest is the ability of the local school district to borrow money (Flanigan, Richardson, & Stollar, 1995). Bonds permit the local school district or the state entity to borrow money to finance long-term building projects, typically for 20 or 30 years (Kowalski, 1989). According to Salmon and Thomas (1981), a bond is a "written financial instrument issued by a corporate body to borrow money with the time and rate of interest, method of principal repayment, and the term of debt clearly expressed" (p. 91). For capital improvements, the issuance of bonds is the most commonly used method for securing funds (Flanigan, Richardson, & Stollar, 1995). School districts are rated in bonding capacity by their records of indebtedness, repayment, assets, and liabilities. The rating report, may either increase or decrease school districts' bonding capacity.

Pay-As-You-Go Funding for Renovation

For the local school district, pay-as-you-go is the most economical method of funding school renovation and maintenance (Flanigan, Richardson, & Stollar, 1995). Most school districts handle maintenance budgets on a regular basis. In economically tight years, funding for school maintenance is meager to ensure adequate reserve for district emergency use. However, most districts do not have adequate resources in the yearly budget to handle large renovation projects (Hartman, 1988). The school district can save substantial interest by paying for the renovation within the current budget year and not issuing bonds (Stollar, 1967). If the total renovation project cannot be funded through the current budget, at least part of the renovation project should be financed as pay-as-you-go (Thompson, Wood, & Honeyman, 1994). Given the prevalent attitude of American society, it may be difficult for the school administrator to sell the community on a slow, gradual, incremental approach to school renovation (Davis & Tyson, 2003). Most often school boards operate as political entities where all sectors of the school district must receive a part of the renovation and maintenance dollars in order to sell the total package (Honeyman, 1998a). Therefore, bonding for capital outlay may be necessary or the best procedure for many small districts and districts that build intermittently (Flanigan, Richardson, & Stollar, 1995).

ALTERNATE SOURCES OF FUNDING FOR SCHOOL MAINTENANCE AND RENOVATION

Local funding options for building renovation and maintenance have both advantages and disadvantages (Alexander & Salmon, 1995). The most common method of funding building renovations is through the issuance of general obligation bonds; however, that is not always possible (Flanigan, Richardson, & Stollar, 1995). Some districts are hesitant to incur additional debt in light of economic conditions (Honeyman, 1994). Other districts cannot issue more bonds until some bonds are retired (Connor, 1998). Therefore, alternatives are often necessary.

State capital bonding is a method used by some states to assist local school districts to fund building projects. Under this plan, states pass large bond issues and supply local school systems with funding based upon need and matching local funding (Hack, Candoli, & Ray, 1998).

If the federal government wanted to make a real difference in education, the most logical area would be in building construction (U.S. Department of Education, 2000; U.S. General Accounting Office, 1996). The federal government has approved federal capital funding, but the consequence has been limited. In fact, most of those dollars are no longer available (Kowalski, 2002).

Some states and local systems used Impacted Area Funding to build schools. By certain state or county or city regulations, developers of substantial projects are required to reserve a specified acreage of land for school construction or are assessed a certain amount of impact fees to cover a portion of school construction and maintenance costs.

School districts in some states are allowed the option of raising a special purpose sales tax in support of school facility construction, renovation, and maintenance. School administrators find it most helpful in generating additional funds for school facility use. Some states even permit school districts to use sales tax dollars to purchase future school sites and pay off school facility indebtedness.

Many states also use a state building authority to construct public buildings, including schools (Kendler, 2003). Chicago, for example, has established a building authority to plan, finance, and build school facilities to lease to the school districts (Hack, Candoli, & Ray, 1998). Some districts have also worked with private sector providers to help alleviate renovation or building needs. For example, a bank in Florida has dedicated extensive space in its facility for the operation of a school to service their employees. Such arrangements provide both the school district and the private sector with advantages that far outweigh any potential disadvantages (Honeyman, 1990).

Other school districts are sharing facilities with other public entities. Some districts make arrangements with public housing authorities, for example, to house early elementary students in the housing authority (Lane & Richardson, 1993). Again, the benefits are great for both groups.

A contractor-financed model of funding for school construction was initiated in South Carolina with considerable success (Chan, 1983). However, school districts need to weigh the advantages and the disadvantages of this model against their particular needs.

Finally, some districts rent facilities rather than build or remodel (Kennedy, 1999). In some rental contracts, a lease-purchase agreement is made to put aside a portion of the monthly rent toward a payment plan of eventually owing the building.

THE BASIS OF BUDGETING FOR SCHOOL RENOVATION AND MAINTENANCE

Many people believe that budgets are about revenues and expenditures. However, the school budget is in reality a three-sided figure, with program being the third side, as illustrated in figure 11.1 (Thompson & Wood, 1998).

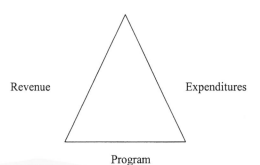

Figure 11.1   Relationship Between Budget and Educational Programs

Programs actually drive both revenues and expenditures (Thompson & Wood, 1998). In schools this is the case particularly in regard to building and renovation. Before the expenditure of funds or requests for funds, the program that will be housed in the educational facility must drive the planning for the construction (Kissing, 1998).

According to Thompson and Wood (1998), budgeting consists of four basic concepts. First, although many professional educators think otherwise, educational programs cost money and funds must be located to support instruction. Second, "the budget is a fiscal expression of the educational philosophy of the district" (Thompson & Wood, 1998, p. 105). Programs that are valued are funded and the district must create some type of priority to determine the value of the programs. Budgeting provides that determination. Third, budgets that establish priorities and plans are created by people who have expertise and influence. Fourth, school budgeting is a political process wherein the public expresses its collective approval or disapproval of the priorities and plans expressed in the budget.

School administrators should have a plan or current cost schedule of materials and labor in place for budgeting the following common renovation and maintenance categories:

1. Plumbing systems
2. Electrical systems
3. Exterior of the buildings
4. Interior of the buildings

5. Structural capacity
6. Special classrooms
7. Heating/cooling system
8. Roofing systems
9. Athletic/extracurricular facilities
10. Library/media center

(Castaldi, 1982; Earthman, 1998; Jarvis, Gentry, & Stephens; 1967; Hartman, 1988; Hawkins & Lilley, 1998; Honeyman, 1998b; Kowalski, 1989; Lackney, 1999; Lane & Richardson, 1993; Ridler & Shockley, 1989).

Cost estimates per unit of construction and maintenance expenditures are available in current building and construction literature by trade and by geographical area. They need to be closely referenced in constructing realistic renovation and maintenance budgets. Many large school districts have in-house architects and engineers to perform accurate budget estimates. Other districts have hired professional construction estimators to perform the task by project or by district.

STRETCHING THE DOLLARS OF THE RENOVATION AND MAINTENANCE PROGRAMS

One of the major problems with school renovation is not the quality of the past construction, but new building codes, requirements, and specifications and the accumulated problems of negligence (Honeyman, 1998b). Since modernization of existing buildings is usually more economical than building new structures, these new requirements constitute a major barrier to school renovation (Henry, 2000). However, some general guidelines may be followed to stretch the dollars for renovation and maintenance programs:

1. Prioritize the elements of renovation and maintenance programs by their essentiality. Some work specified by codes has to be performed to get the inspectors' approval.
2. Leave as much of the building structure as possible untouched unless it is absolutely necessary. Structural changes are costly.
3. Keep change orders to a minimum because they are both costly and burdensome.

4. Evaluate your budget from time to time to get a good feel of where it stands and what else can be afforded.
5. Price comparisons of alternative approaches is crucial in making renovation and maintenance decisions.
6. Stocking maintenance items at low prices and bidding in large volumes achieves the best economy of scale.

In addition, recent economic situations also demanded that states use generic building programs to help fund facilities (General Accounting Office, 1995; Geiger, 2001; Jones, 2002). States building 50 schools can get better prices than a district building one school (Jarvis, Gentry, & Stephens, 1967; Kendler, 2003; Kowalski, 1989; Wood, 1986). As a result of this generic approach, more tax dollars are saved for other school renovation and maintenance projects.

SUMMARY

This chapter explores the many possible options for the funding of school renovation and maintenance projects. Most school maintenance programs are funded out of the school district general fund. Major school renovation projects are funded either pay-as-you-go or by issuing bonds amortized over a period of 20 or 30 years. Bonds provide the local school district with a method of borrowing money. The bond is a contract between the bondholder and the district, which promises to repay the bond principal at maturity and the accrued interest. More school districts are considering alternative methods for the funding of school renovation and maintenance projects. Sales tax increases have been gaining more acceptance as a popular alternative method to meet school facility needs.

EXERCISES

1. Select a district that is anticipating a major school renovation effort and ask the superintendent, school business manager, or school board president the following questions:

   A. What are the avenues for funding school renovation projects?
   B. How will the renovation be financed? Will it be through bonded indebtedness or pay-as-you-go?

   C. Is the local tax base sufficient to support the renovation?

   D. Will capital outlay expenditures reduce the finances available for current operating costs?

   E. Could at least part of the capital outlay program be financed on a pay-as-you-go basis?

2. Select a district that just completed a major renovation project and ask the superintendent the above questions.

3. Determine if there is a difference between the information provided by the two different districts. What could they learn from each other?

4. Locate a bond attorney and interview him or her. Ask him or her to describe the process of selling municipal bonds for local school districts.

5. Select a local school district that recently issued school bonds and ask to examine the bonds.

6. Discuss with the finance director of your school district the possibility of raising the sales tax in support of school facility projects. What procedures do you need to go through?

12

EVALUATION OF RENOVATION AND MAINTENANCE PROGRAMS

SCENARIO

M r. James Bennett, superintendent of the Washington County School District, is a believer in saving existing school buildings. For the past three years, he was able to convince the school board to direct a large amount of money toward school maintenance and renovation. School facilities in Washington County are well maintained and five older school buildings have been renovated since he started his superintendency. As a matter of fact, funding that was originally allocated for the construction of a new elementary school was used to completely renovate one existing elementary school and two old elementary schools that were closed a year ago.

In preparing next year's budget, Mr. Bennett understands the tight economic situation statewide. It is not easy for the school board to approve additional dollars for facility renovation and maintenance. Some school board members are thinking of cutting a maintenance budget that is excessive to them. Some are leaning toward postponing capital outlay programs, including renovation projects, as a means to balance the budget. What Mr. Bennett wants to see is the protection of the school maintenance budget and an uninterrupted schedule of school renovation. He knows that he has to bring up solid facility evidence to justify to the school board that school renovation and maintenance programs have been heading the right direction in the past three years. To collect data about the effectiveness of the school renovation and maintenance programs, Mr. Bennett has asked Mr. David Jones, the school facilities director, to design and conduct a school renovation and maintenance evaluation. As a school renovation and maintenance evaluation has never been performed in the Washington County School District before, Mr. Bennett has approved the use of an outside facility consultant to assist Mr. Jones in conducting the evaluation. Since time is of the essence, Mr. Bennett has instructed that a qualified facility consultant be brought in immediately to start planning for the evaluation activities of the school renovation and maintenance programs.

RATIONALES FOR EVALUATION PROGRAMS

Many school districts are so busy with their overall facility planning work that evaluation of school renovation and maintenance programs is often neglected. The same is true of new school construction. Planners, for one reason or another, seldom go back to newly constructed buildings to evaluate the process and the outcome of their planning efforts. However, this does not mean that evaluation is not needed. Evaluation of school renovation and maintenance programs provides a basis for program improvement. Specifically, the rationales for evaluating school renovation programs are:

- To measure the extent to which the renovation goals are achieved.
- To evaluate if the right decisions were made during renovation.
- To assess the effect of the renovation project on overall facility needs.
- To review if the renovation priority is still justifiable.
- To examine the degree to which the renovation project is in compliance with the educational specifications.
- To summarize how the school building was brought up to current codes.
- To evaluate if the planning approach was appropriate.
- To measure the degree to which the education process was interrupted.
- To investigate how well the renovated school building accommodates the education program (Macclay & Earthman, 1992).

Architects and engineers involved in a school renovation project may choose to have their own evaluation approach that will be focused more on the design aspects. One important rationale of their evaluation is to examine the overall effectiveness of the renovated facility. The conditions before and after renovation will be compared. Designers will also determine cost efficiency as a justifiable rationale for project evaluation.

An annual evaluation is necessary for a school maintenance program to justify the budget. Strong rationales can be drawn to support the evaluation of school maintenance programs as follows:

- To assess if the maintenance needs of the year have been met.
- To demonstrate the effectiveness of the program in maintaining school facilities.
- To evaluate the human resources allocated to the operation of the maintenance program.
- To examine the cost efficiency of the components of the maintenance program.
- To justify the priority list of preventive maintenance items.
- To evaluate the quality of all contract services.
- To examine the efficiency of service delivery in response to school maintenance requests.

The school maintenance budget needs to be well protected from cuts, particularly in times of economic difficulty. A sensible rationale will add strength to support for school maintenance.

INVOLVEMENT OF EVALUATION PROGRAMS

For an evaluation of school renovation and maintenance programs to be effective, three elements of involvement should be considered: the extent, the time, and the people.

The Extent

The evaluation of school renovation programs should cover both the school district-level plan and the individual school renovation projects. Evaluation of the school district-level plan should focus on the justification of all the proposed school renovation projects and their priorities. Individual school renovation projects can be evaluated with emphasis on input, process, and output of the renovation projects. For school maintenance, evaluation can also be conducted both at the district and the school levels. Evaluation of school maintenance at the school district level should include a review of district planning efforts and the overall achievement of these efforts. Evaluation of maintenance programs at the school level should focus on physical improvements as a result of program effectiveness.

The Time

School renovation planning and school maintenance planning at the district level should be reviewed every year so that effective work can be identified and reinforced, whereas ineffective components can be studied for improvement. Maintenance work at the school level should be evaluated annually to ensure that the school continues to be maintained in a healthy and clean manner. School renovation projects should be evaluated one year after substantial completion so that occupants can provide feedbacks from their experience of using the renovated facility. If evaluation takes a comparative approach, then data about the building conditions need to be collected before and after renovation.

The People

Evaluation of school renovation and maintenance programs should take a participatory approach to avoid subjectivity. In general, the planning committee that initiated the program should be responsible for program evaluation. But in some school districts, independent committees are assigned to perform the program evaluation task. For school renovation and maintenance programs at the district level, evaluation committees may consist of the district facility planning director, maintenance director, assistant superintendent of business operations, school principal representatives, teacher representatives, and community representatives. Members of the school maintenance evaluation committees may include school principals, the district maintenance director or his designee, teacher representatives, community representatives, and custodial representatives. A balance between internal and external representatives should be kept to maintain objectivity. Committees to evaluate individual school renovation projects may consist of the school district facility planning director, the maintenance director, the school principal, teacher representatives, the head custodian, architects, and engineers. The key to the success of these committees is to draw as many members as possible from the original planning committee who have a thorough understanding of the entire planning and renovation process of the project.

APPROACH TO EVALUATING MAINTENANCE AND RENOVATION PROGRAMS

The evaluation of school renovation and maintenance programs can take many approaches. Some of the common approaches are: input-process-output approach, criterion-referenced approach, comparative approach, and goal attainment approach. It is not uncommon to see school districts employing a combination of these approaches to evaluate their school renovation and maintenance programs.

Input-Process-Output Approach

The input-process-output approach evaluates a program in three phases: the input, the process, and the output. Evaluation on the input reviews the human and fiscal resources allocated to support the program. The process evaluation examines every phase of the process to see if school renovation and maintenance programs are handled appropriately. The outcome evaluation focuses on the end product of school renovation or maintenance programs. An advantage of the input-process-output approach is that it allows evaluators to be exposed to all phases of a renovation or maintenance program.

Criterion-Referenced Approach

Program evaluation taking a criterion-referenced approach uses a set of predetermined criteria to assess program effectiveness. Examples of some common criteria are:

- All restroom facilities must be renovated to meet the current disability codes.

- Computer systems must be rewired to accommodate Internet access in every classroom.
- All classrooms must be designed to accommodate a student body according to the new state class-size regulations.
- All outside school landscaping must be attractively maintained.

School districts evaluating school renovation and maintenance programs work with these criteria to ensure they are met. An advantage of the criterion-referenced approach is that it provides specific evaluation guidelines that are understood by all parties throughout the entire school renovation or maintenance program.

Comparative Approach

A comparative approach to evaluation of school renovation and maintenance programs takes two formats. One is the before-and-after format that takes into account the physical conditions of school buildings before and after program implementation. The effectiveness of the program is measured by the differences between the data before and after renovation or maintenance. The other evaluation format is the side-by-side format that compares the performance of one program with that of a similar program. Comparative approach as a method of program evaluation provides evaluators with the option of examining other programs or projects similar to the ones under evaluation.

Goal Attainment Approach

The goal attainment approach takes into consideration the extent of fulfillment of the renovation and maintenance goals that were created when the project was initiated. The program is considered highly successful when all the goals are fully satisfied. An advantage of the goal attainment approach is that no one can dispute goal attainment as a measure of accomplishment.

In considering the four evaluation approaches, the goal attainment approach is the most convenient and time efficient, since goals were clearly established for each program at the beginning. The comparative approach is time consuming and seldom used because the results of the comparisons are of interest to only a few groups. Both the input-process-output approach and the criterion-referenced approach pay a great deal of attention to program details as foci of evaluation. What school districts usually do is to take advantage of the best of these approaches to tailor an approach that generates the kind of data that fit their particular needs.

PROCEDURES FOR EVALUATING MAINTENANCE AND RENOVATION PROGRAMS

The evaluation of renovation and maintenance programs differs from district to district. Some school districts do not even require a formal evaluation of these programs. For those that require an evaluation, evaluating a renovation program could be completely different from evaluating a maintenance program. School districts usually tailor their evaluation methodologies to meet their particular needs. However, certain procedures can be followed.

Before Program Implementation

This method of evaluation is customarily a component of program development. Before the program is implemented, the planning committee should discuss and decided on an agreeable method of evaluating the program. This will allow enough time for the program administrators to understand what will be evaluated so that they will be prepared for the terminal evaluation. If the methodology calls for a comparison of before and after effects, then data concerning the building conditions before the program or project was implemented should be collected at this stage.

During Program Implementation

Some school districts call for interim evaluation of their renovation and maintenance programs. Interim evaluation reports will indicate if the programs are on the right track. Even though interim evaluation is not always required, program administrators need to get ready for the program-end evaluation by logging daily events and special issues, with full documentation. School boards and superintendents need to be updated periodically about the progress of the programs.

After Program Implementation

If a program calls for only one evaluation, the evaluation is always placed at the end of program completion. This makes good sense because the end of the program is an excellent time to review the overall affect of program implementation on the school buildings. Evaluation at the end of the program should be based on the agreed methodology as described at the beginning of the program. Sometimes an independent committee may be organized to evaluate the outcomes of a renovation or maintenance program. The end-of-program evaluation should review the implementation in addition to the results of the programs.

SCOPE FOR EVALUATING MAINTENANCE
AND RENOVATION PROGRAMS

Evaluation of school renovation and maintenance programs can cover a great deal of territory. However, the basic question always goes back to the original purposes of initiating the program. The scope of the evaluation used by one school district may not apply to another because the purposes of their programs are different. The following lists of evaluation items are representative of the potential scope of program evaluation.

Evaluation of school renovation programs at the district level could include the following areas:

- Identification of schools for renovation
- Justification of school renovation needs
- Cost estimates of school renovation projects
- Priorities of school renovation projects
- Consolidation of renovation projects into district capital outlay program
- Alternative funding for school renovation program

Evaluation of individual school renovation projects can be very involved. However, the fundamental elements are:

- Planning participation
- Planning process
- Compromising decisions
- Coordination effort
- Cost efficiency
- Budget control
- Meeting schedules
- Quality control

- Compliance with new codes
- Meeting requirements of educational specifications
- Functional effectiveness
- Aesthetics
- Operation of new building systems
- Extension of school building life

The school maintenance program at the district level is sophisticated. Evaluation of the program should include the following areas:

- Goal setting of the program
- Preventive maintenance effort
- Efficiency and effectiveness of maintenance department organization
- Appropriateness of budgeting process
- Operational procedures of the maintenance department
- Utilization of maintenance resources
- Outsourcing decisions
- Responses to emergencies
- Adequacy of allocation formulae
- Supervision of maintenance quality
- Development of maintenance staff
- Participation in new facility planning
- Systematic archiving of school building information
- Support of school custodial services
- Effort toward a safe school environment
- Initiatives toward energy conservation

Evaluation of a maintenance program at the school level is essential because the effectiveness of the program can easily be detected by the community. The following areas should be included in the scope of evaluation:

- Identification of custodial services
- Staffing of custodial crew
- Scheduling of custodians
- Supervision of custodial work
- Working toward a safe school
- Energy conservation effort

- Cleanliness and healthiness
- Community support
- Aesthetics
- Staff development
- Periodic system checking and servicing
- Responses to school emergencies
- Budget control
- Assistance to faculty

Listed above are some of the general areas for the evaluation of school renovation and maintenance programs. Special features of individual school districts are not included. For example, recent school security programs include the use of surveillance systems and security officers. They need to be included as part of the school renovation and maintenance evaluation.

STAKEHOLDERS' PERCEPTION OF PROGRAM EVALUATION

Stakeholders involved in the school renovation program should be surveyed or interviewed so that their perceptions of school renovation can be systematically analyzed. These stakeholders are the district facility planner, the architect, the engineers, the school principal, the teachers, the community representatives, and the school staff. They went through different phases of school renovation together and can help improve the process of school renovation by providing valuable feedback. Among the stakeholders, the occupants of the renovated school in particular can furnish a great deal of information about the effectiveness of the "reborn" building. Facility planners can learn a valuable lesson from what they experienced.

Most school visitors are polite and do not usually express anything negative about the maintenance of the school. People who can give an honest opinion about the upkeep of school buildings are the district custodial supervisor, the school custodians, the faculty and staff, the parents, the students, and community members. An example of surveying the outlook of school buildings (Strickland & Chan, 2002) is included in appendix H. Surveys of the perceptions of these stakeholders can generate useful data for the improvement of school maintenance programs. In fact, an immediate effect of surveying students about school building upkeep is to foster their concern for the school building to which they belong. This will in turn help reduce vandalism to the school.

USE OF PROJECT EVALUATION RESULTS

Since the school board approves facility projects, project evaluations need to be reported to the school board for review (White, 1992). Copies of the evaluation results should be available to all the stakeholders for reference. Significant results of project evaluation include:

- Improving the facility planning process;
- Determining the resource input of future projects;
- Better preparing budget estimates for future projects;
- Documenting the effectiveness of the renovated building systems;
- Testing the durability of building materials;
- Verifying the practicality of the redesigned spaces;
- Ensuring all current building, fire, and disability codes are met;
- Examining the improvement of school building safety;
- Enhancing the efficiency of project scheduling;
- Determining the choice of equipment for future projects;
- Deciding on the need for frequency of future services;
- Confirming the effectiveness of energy conservation designs; and
- Justifying outsourcing decisions.

The results of project evaluation also can be used to serve other purposes. Facility planners just have to design an instrument to generate the needed data for their use. For example, to solicit the building occupants' perception of the outlook of a renovated school, a special evaluation instrument could be constructed to include the school's front entrance, the new brick veneer, the new walkway, the new canopy, the new landscape, the new driveways, the new school sign, and the new parking area.

SUMMARY

This chapter covers many aspects of the evaluation of school renovation and maintenance programs. It calls the attention of school administrators to school facility program evaluation to generate useful data for improvement. Evaluation of school renovation and maintenance programs should cover the essential areas at both the school and the district levels. Program evaluation of school renovation and maintenance is a significant step in facility planning and management. Procrastination because of administrators' heavy workload is not an acceptable excuse. For new facility programs to be planned successfully, current or previous facility programs need to be evaluated for efficiency and effectiveness.

EXERCISES

1. Survey the school administrators responsible for the facilities in several school districts of various sizes in your state. Check how many of them conduct evaluations of school renovation projects. If not, why not?

2. What is the merit of evaluating the school renovation program at the district level? How could results of the evaluation be used to improve renovation programs?

3. How can a districtwide facility maintenance program be operated efficiently and effectively? Interview two directors of school maintenance to get insight into the operation of the program.

4. Construct a checklist of existing school buildings with an emphasis on cleanliness and healthiness. Field test it at several schools and revise the checklist afterward if necessary.

5. Design an evaluation project to justify the large amount of the school maintenance budget.

6. Check with two other school districts in your neighborhood and explore how school renovation and maintenance projects are managed.

7. In school facility planning, some school board members support school renovation projects while some prefer to construct new school buildings. As director of school facilities, prepare a 10-minute presentation to the school board meeting promoting the benefits of school renovation. In the presentation, include data generated from the most recent renovation to support your position.

8. Compare two school renovation evaluation instruments. Identify their foci of evaluation, discuss their strengths and weaknesses, and comment on their validity.

SCHOOL MAINTENANCE AND RENOVATION: A FUTURE PERSPECTIVE

SCENARIO

The purpose of school maintenance is to ensure that a school building will last longer to serve its educational use. Renovation is needed to bring old facilities up to modern program requirements. This makes good sense in time of budget constraints. Mrs. Armstrong, superintendent of Kirby County schools, just came back from attending a state workshop on school facilities that focused on school maintenance and renovation issues. Mrs. Armstrong is impressed with the idea that proper school maintenance and renovation could result in constructing fewer new school buildings. She realizes that the future does not look promising in view of the state's economic situation. Therefore, the time has come for the school district to come up with a plan to help itself. Even though Mrs. Armstrong is convinced that she would support any effort toward increased attention to school maintenance and renovation in the years ahead, she has no idea where to start initiating a program that would cover effective school maintenance and renovation. She has asked the director of school planning and director of school maintenance to meet with her next week to discuss these matters. Meanwhile, she is organizing the materials she brought back from the state workshop and asks her secretary to make copies for distribution to the directors to read before they come for the meeting next week. She is really looking forward to the feedback from her directors.

SCHOOL FACILITIES AS A SUPPORT FOR EDUCATIONAL PROGRAMS

Much has been said about school facilities as a support for educational programs. Not only does a school building provide the physical space that is needed in the operation of educational programs, it is a visible system of decision makers' support for education. A well-designed school building supports the educational program from both the inside and outside. It incorporates the many brilliant ideas from subject area consultants, teachers, staff, administrators, parents, students, community leaders, planners, and designers. Careful examination is made of the day-to-day teaching process to achieve instructional efficiency and effectiveness. Building design features are focused on physical improvements to facilitate better student learning outcomes. Attention is also given to aesthetics in designing facilities to achieve an environment conducive to learning. Well-designed school buildings need to be well maintained to serve their design purpose. They need to be renovated after a number of years to maintain their efficiency and effectiveness.

PROGRAM CHANGES AND FACILITY ACCOMMODATION

Educational programs change in one or more of these aspects:

1. Change in the academic content of the discipline;
2. Creation of new programs;
3. Celetion of existing programs;
4. Change in program delivery approach; and
5. Change in school organization that affects the program.

Since school facilities are designed to support educational programs, any changes in educational programs mean that facilities need to be changed to continue their support of education. For example, a laboratory facility has to be set up to accommodate a newly added dissection activity in biology study; space has to be modified to house new counseling programs in elementary schools; closing of the woodshop program means additional space that can be converted for other program use; the change to a CAD system for graphic design means changing the drafting table setup to a computerized workroom; the change of a junior high school to a middle school involves a major configuration of the space in a school building.

A school building undergoes many physical changes in its life to accommodate the changes in educational programs. Staff from the school district maintenance department usually perform minor building changes such as addition of electrical outlets, erection of simple partition walls, and the replacement of doors and locks. Major changes include upsizing the HVAC unit, removing load-bearing walls, and upgrading the existing safety alarm or sprinkler system. Since these changes may affect compliance with building and fire codes, they are considered renovation activities. In this case, an architect is usually involved to design the renovation project to ensure that the building is properly converted for its newly assigned use.

SCHOOL MAINTENANCE AND BUILDING CONDITIONS

New school buildings start to deteriorate from the first day of operation. Some deteriorate quickly because of poor construction quality or poor maintenance. For various reasons, mainly because of budget and time constraints, some school buildings were designed and built with the minimum quality that would meet building and fire codes. Specification of poor quality materials, together with poor workmanship, contribute to new school buildings that are either expensive to maintain or that deteriorate quickly. On the other hand, quality school buildings need to be well maintained to last. Poor maintenance of school buildings, regardless of building quality, simply accelerates the building deterioration process. This explains why some school buildings last for more than 60 years while some have to be abandoned before 40 years of use.

Another reason for building deterioration comes from misplaced priorities in renovation projects. As stated in previous chapters, the purposes of school renovation are to improve the outlook of the facility, to bring the facility up to the current codes, and to address issues of building safety. However, in some renovation projects, renovation money is spent upgrading the building outlook at the expense of building safety and steps to maintain and prolong the healthy life of a school building. Consequently, unsafe school buildings remain unsafe despite their attractive outlook.

B. H. Krysiak's observation of deferred maintenance, summarized by Kowalski (2002), really hits the point:

> Often inexpensive facilities are constructed on the premise that the next generation of taxpayers will pay to improve or replace them. But when the next generation abdicates this responsibility, an even greater burden gets passed to a third generation of taxpayers. This philosophy and an overall neglect for proper maintenance have certainly contributed to the facility crisis today. (p. 63)

THE TREND TOWARD EXPANDING SCHOOL RENOVATION AND MAINTENANCE PROGRAMS

It is clear to educational decision makers that school enrollment will keep growing, the funding in support of education, particularly of school construction, does not look promising, and the need for additional classrooms is compounded by class-size reduction mandates in many states. Some school districts resolve their classroom shortage problem by hauling in portable classrooms, while some tackle the problem by building new additions to existing school buildings. When portable classrooms provide only a temporary solution and increasing the property tax rate does not seem welcomed, more and more school districts will be paying close attention to increasing support for school maintenance and renovation projects. The simple reason is that efforts in maintenance and renovation will prolong the useful life of school buildings. In an age of pressing needs for additional classrooms and tight revenue situations, school districts' response of increasing school maintenance budgets and initiating building renovation projects has long-term benefits. Because of attention to maintenance, school buildings can last longer. Because of renovation work, many obsolete school buildings can be restored to active use. Consequently, more tax dollars can be saved. The mathematics is simple to understand. If a school building can last an average of 15 more years beyond normal life expectancy, the school district will have a net gain of 15 years of use and a replacement school will not be needed for another 15 years. If every school building in the school district can last for 15 more years, the savings to the school district are enormous. There is no reason why school boards and the public would not support increased school maintenance and renovation budgets.

A RENOVATION AND MAINTENANCE PLANNING MODEL

When a school district is used to planning a tight budget, a proposal to increase the school maintenance budget will not pass without good justification. This is because most budget planners focus their attention on balancing the budget for the upcoming year, while overlooking the long-term savings of school maintenance work. Equally difficult is a proposal to renovate obsolete classrooms to save educational funding that would otherwise be used to construct new school buildings. Communities promised new school buildings are often disappointed to find that their long awaited new facility turns out to be a renovated school. Consequently, school districts find it very difficult to make the paradigm shift from the concept of cutting maintenance budgets to increasing maintenance budget, and from promoting new school facilities to favoring school renovation. Not only do school administrators have to handle the politics that come with these decisions, but also they need to produce sensible evidence to justify their recommendations. The following is a description of the major components of a proposed planning model (see appendix F) that provides a step-by-step approach to systematically collect and analyze data to produce a cost comparison of alternate facility plans for public review:

Enrollment Forecast. Student enrollment in a school district needs to be closely monitored by accurately performing 1-year, 2-year, 5-year, and 10-year forecasts.

Facility Projection. School facility planners need to transfer forecast student enrollment into classroom needs by grade and by school level, taking into account the school locations and the growth factors of communities.

Facility Inventory. A complete inventory of school facilities by school level needs to be conducted with a focus on standard classroom counts and operational realities.

Identification of Facility Needs. The net facility needs can be estimated by comparing the projected facility needs and the inventoried facilities of a school district.

Alternatives of Meeting the Facility Needs. Facility needs of a school district
can be met by building new schools to meet the growth and replacement
needs, by consolidating schools to make best use of facilities, by closing
obsolete schools, and by reworking the configuration of school organization to fully utilize existing facilities.

Facility Plan for Implementation. After the decision is made on how the facility needs can be met, a facility implementation plan can be developed.
A school district facility plan typically includes new construction projects, renovations projects, and the facility maintenance program. A time
factor is usually installed to indicate project priority.

Evaluation of Facility Plan. The purpose of the evaluation function is to examine the existing district facility plan to consider the possibility of inte-
grating an expanded school maintenance and renovation (ESMR) program.
It is anticipated that an ESMR program will result in downsizing the scope
of major new construction efforts, thus saving substantial tax dollars for
school district use.

Cost Comparison. The cost of the existing facility program is compared with
the cost of the alternative program that incorporates the ESMR. It is anticipated that the alternative program will be more economical in the long
run.

The purpose of this maintenance and renovation planning model is to provide a step-by-step approach for facility planners to carefully examine the facility needs of their school district. It uses a simulation method by evaluating
existing facilities and offering alternatives that can be pursued to fully utilize
existing facilities. Ideally, the cost estimate tied to each component of the program will provide strong evidence to convince the school board and the public that ESMR will work.

A LOOK TO THE FUTURE: CHOICES

The need for additional classrooms arises as a result of student population
growth, class-size reduction mandates, and the closing of outdated facilities.
However, when tax dollars are tight, school districts cannot afford to meet all
their classroom needs by building new classrooms. Renovating old school
buildings makes sense because the same number of classrooms can be remod-

eled with much less money than the cost of constructing an equal number. Additionally, an excellent maintenance program will keep the school in operation for many more years, cutting down the frequency of having to build new schools. The choices are between paying more for maintenance and renovation upfront to save in the long run, or staying with minimum maintenance and renovation budgets, which will result in more expensive school construction in the future. The first choice obviously makes sense in the long run. However, some school districts have serious budget problems. They survive by squeezing a few dollars from account to account and are not in a climate to push for expanded maintenance and renovation programs. When the school districts' financial situation improves, they need to start thinking of reserving money for an expanded maintenance and renovation program.

SUMMARY

This chapter summarizes the main theme of this book: Maintenance and renovation are the key to school facility planning. When all the cost estimates are disclosed, the choice of long-term saving is clear. School boards may elect to pursue a different direction in facility planning for various reasons. The facts, however, speak for themselves. Expanded school maintenance and renovation programs represent a worthwhile investment for school districts in the future.

EXERCISES

1. Based on the scenario at the beginning of this chapter, develop an alternative facility plan to include the expanded school maintenance and renovation program for Mrs. Armstrong's review. Highlight your alternative plan and summarize the selling points to the school board and the public.

2. Develop a detailed plan of how you can evaluate the existing maintenance program and how you can identify areas of improvement in terms of facility needs.

3. Prepare a short speech to the school board when presenting your alternative school facility plan. Materials in your speech should cover the background of this alternative plan, the current issues of facility planning, and the long-term benefits to the school district.

4. Locate a copy of the long-term facility plan in your school district. Study the background of its development and design two alternative plans to reflect the implementation of the expanded school maintenance and renovation program.

5. Southside High School is 45 years old. It was declared obsolete because many building systems need to be totally replaced. The decision was made five years ago to replace Southside High School with a new school building. Now, with the new concept of expanded school maintenance and renovation programs, school administrators are taking a second look at the school building and think that it could be extensively renovated and used for another 20 years. Thus, the recommendation is made to revise the school facility plan to renovate Southside High School instead of replacing it. You are asked to make a presentation in a public hearing to discuss the issues involved in this change of plan by disclosing building data to support your recommendation.

6. Given an old school building to be renovated, how do you come up with a realistic cost estimate to convince the public that the building can be renovated for substantially less than building a new school of similar size? Where do you draw the line at construction costs? What factors do you need to take into consideration?

7. When a districtwide school facility plan has been developed and approved by the school board, what justifications do you have as a school administrator to revisit the plan development and recommend an alternative plan? What channels do you need to go through? Will the state department of education be involved?

APPENDIX A

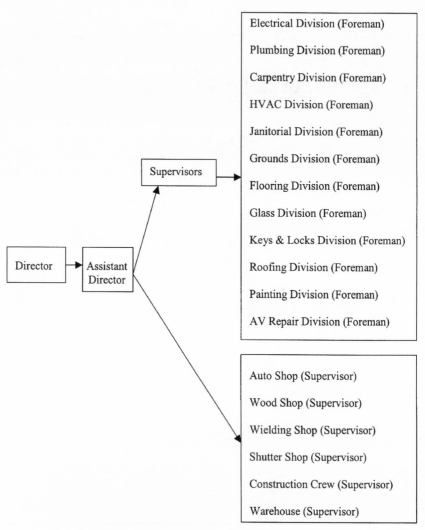

Figure A.1.   Organization of a School District Maintenance Department

APPENDIX B

SCHOOL MAINTENANCE CHECKLIST

Interior of School Building

Daily Checking:

_____ Fire alarm system
_____ Intrusion alarm system
_____ Surveillance system
_____ Public announcement system
_____ School bell
_____ Central controlled clocks
_____ Emergency lights
_____ Backup battery
_____ Telephone system
_____ HVAC system
_____ Exit signs
_____ Fire hoses
_____ Fire extinguishers

Periodic Checking:

_____ Sprinkler system
_____ Heat detectors
_____ Smoke detectors
_____ Fire door operation
_____ Gym bleachers
_____ Gym equipment
_____ Theater equipment
_____ Classroom furniture
_____ Door and window locks
_____ AV equipment
_____ Custodial equipment
_____ Wall conditions
_____ Ceiling conditions

_____ Exit door panic hardware
_____ Emergency shutoff valves
_____ Emergency windows
_____ Obstruction free hallways
_____ Emergency eye washer
_____ Emergency shower
_____ Fume chamber
_____ Chemical storage
_____ Mechanical exhaust systems
_____ Restroom water flushing
_____ Restroom cleanliness
_____ Kitchen equipment temperatures
_____ Boiler water temperature
_____ Light fixtures
_____ Drinking fountain cleanliness
_____ Floor conditions
_____ Restroom vandalism

_____ Window conditions
_____ Venetian blinds

EXTERIOR OF SCHOOL BUILDING

Daily Checking:
_____ Debris or garbage

_____ Walkways and driveways
_____ Water drainage
_____ Dumpster area
_____ Security lights
_____ Parking lot lights
_____ Playground equipment
_____ Exterior walls
_____ Portable classroom conditions

_____ Outside storage
_____ Flag poles
_____ Message on announcement sign

Periodic Checking:
_____ Gutters, downspouts, drains
_____ Plants and grass
_____ Air-conditioning tower
_____ Stadium bleachers
_____ Roof hatches
_____ Grease trap
_____ Retention pond
_____ Roofing conditions
_____ Exterior window caulking
_____ Air-cooled condenser
_____ Catch basin
_____ Field house

____ School sign
____ Front entrance
____ Athletic fields

Remarks:

Checked by: _____
Date: _____
c.c.   School Principal
         Head Custodian

APPENDIX C

Table C.1. A Progress Chart for School Renovation

Categories	Get ready	Wk1	Wk2	Wk3	Wk4	Wk5	Wk6	Wk7	Wk8	Wk9	Wk10
Secure permits	X										
Secure bonds											
Secure insurance	X										
Check licenses	X										
Order materials	X										
Order equipments	X										
Schedule workers	X										
Remove school furniture and equipment	X										
Demolition		X	X								
Clean up site debris		X	X								
Mechanical work		X	X	X	X	X	X	X	X	X	X
Structural repair		X	X	X				X	X	X	
Construct new walls		X									
Electrical work			X	X	X	X	X	X	X	X	X
Plumbing work			X	X	X	X	X	X	X	X	X
Install new windows					X	X	X				

Task	1	2	3	4	5	6	7	8
Install sheet rock	X							
Install ceiling grids		X	X					
Install new doors		X	X					
Paint			X	X	X	X		X
Install floor tiles			X	X	X	X		
Install new cabinets				X	X	X		X
Install new carpet					X	X		
Install new bulletin and chalkboards					X			
Install new door locks				X		X		
Install new ceiling tiles						X	X	
Building inspection	X			X				
Fire inspection								
Health inspection								
Check all systems				X		X		
Deliver new furniture				X		X		
Clean up school				X		X		
Renovation completes and staff moves in						X		X

APPENDIX D

Table D.1. Larks County School District Fifth-Year Projection of Classroom Needs

School	No. of Clssrm	Avg. Class Size	Current Enrllmt	Current Clssrm Needs	Current Classrm Shortage	5th-Yr Enrllmt	5th-Yr Classrm Needs	5th-Yr Classrm Shortage
Brown Elem.	28	18	410	23	0	505	28	0
Willy Elem.	30	18	435	25	0	548	31	1
Bryant Elem.	30	18	526	30	0	641	36	6
Wade Middle	45	22	1028	47	2	1365	62	17
Lark High	50	25	1248	50	0	1561	63	13
Total	183	n/a	3647	175	2	4620	220	37

APPENDIX E

Table E.1. Buford County School District Priority List of Capital Outlay Projects, Revised May 24, 2003

Priority	Capital Outlay Projects	Projected Costs ($)
1	New elementary school construction on Bakers Chapel Road	4,500,000
2	New elementary school construction on Glenwood Road	4,500,000
3	Renovation of Buford County Middle School	1,200,000
4	Addition to Brunswick Middle School	1,800,000
5	Renovation of Brooksfield Elementary School	1,000,000
6	New Browns High School Construction on Tinkersley Road	12,100,000
7	Conversion of old Browns High School to Browns Middle School	2,500,000
8	New elementary school construction on Douglasville Highway	4,500,000
9	Renovation of science laboratories in Dickerson High School	1,000,000
10	Site acquisition for a new middle school at Dunkin-Blue Ridge Area.	300,000
11	New P.E. Building for Clarksville Elementary School	250,000

(continued)

Table E.1. *(continued)*

Priority	Capital Outlay Projects	Projected Costs ($)
12	New P.E. Building for Long Elementary School	250,000
13	Construction of new Buford County Middle School on Kingston Road	8,000,000
14	Renovation of Edison Elementary School	1,000,000
15	Addition of performance art building to Dickerson High School	2,000,000
16	Addition to Hollingsworth Middle School	3,000,000

APPENDIX F

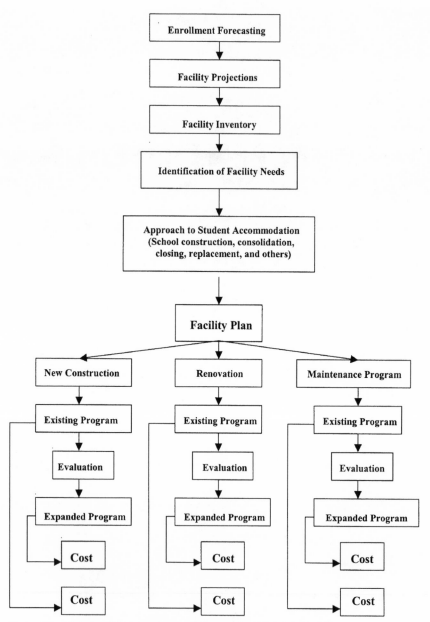

Figure F.1. A School Maintenance and Renovation Planning Model

APPENDIX G

WHAT SCHOOL ADMINISTRATORS MUST KNOW ABOUT FACILITY PLANNING AND MANAGEMENT

1. Personnel:

   A. Assign a district level administrator to be responsible for handling school facility business regardless of the district size.
   B. Create a clerk-of-the-work position to supervise the technical aspect of school construction.

2. Database:

   A. Perform accurate student enrollment projections to assess the direction and magnitude of district growth.
   B. Conduct an inventory of school facilities in the district to record space availability and needs.

3. Planning Options:

A. Explore all possible options for accommodating students such as:

- Constructing new school buildings
- Renovating old school buildings
- Adding new wings to existing school buildings
- Consolidating existing schools
- Closing inefficient school buildings
- Reconstructing school grade organization and location
- Planning new school attendance boundaries

B. Fully evaluate all the available options and make recommendations to address the student population issue.

4. Planning Process:

A. Check and comply with all federal and state mandates.
B. Involve all stakeholders in a participatory planning process.
C. Pay attention to details in every phase of planning and construction.

5. Public Relations:

A. Keep the public informed and involved at appropriate phases of planning and construction.
B. Win public support by planning functional and attractive school buildings.

6. Maintenance:

A. Explore alternatives to save in the daily operation of schools.
B. Defend your school district's maintenance budget. Develop a preventive maintenance program for school.

APPENDIX H

SCHOOL BUILDING/CAMPUS "CURBSIDE CRITIQUE"

Please critique the campus to assess your critical "curbside image." Your curbside image is the impression given when community members pass by or come to your school. Some people will never enter your front door; many times their only impression of your school is "just passing by." Unkempt building and grounds can give a negative impression of the care you give the taxpayers' property and even the children with whom you are entrusted.

SCHOOL/CAMPUS _____

DATE _____

	YES	NO
1. Is the parking lot clean and free of debris?	_____	_____
2. Are windows clean (void of dried-up tape)?	_____	_____
3. Are the grounds and sidewalk free of weeds, cigarette butts, scattered trash?	_____	_____

4. Is the front door clean (void of children's _____   _____
   fingerprints, freshly painted)?

5. Are the items displayed in windows and on doors _____   _____
   current with the season?

6. Is the trash area clean (without overflowing _____   _____
   dumpsters)?

7. Are the gutters and downspouts in good conditon? _____   _____

8. Are the draperies/blinds neat and _____   _____
   organized-looking?

9. Are the flowerbeds free of weeds? Sidewalks? _____   _____

10. Is the lawn mowed and well manicured? _____   _____

11. Are the walls properly maintained with no graffiti? _____   _____

12. Is playground equipment safe? _____   _____

13. Are entrances attractively landscaped? _____   _____

14. Are all exterior lights burning? _____   _____

15. Are parking lots and playgrounds free of puddles? _____   _____

16. Is exterior paint adhering properly? _____   _____

Comments for Maintenance: Record any condition that threatens safety (i.e., a broken step, pothole, faulty railing, etc.).

_____

_____

_____

_____

_____

Signature
Cc: Maintenance Director
School Principal

The "School Building/Campus Curbside Critique" form is obtained with permission of authors from: J. S. Strickland & T. C. Chan. (2002). Curbside critique: A technique to maintain a positive school yard image. *School Business Affairs*, *68*(5), 24–27.

APPENDIX I

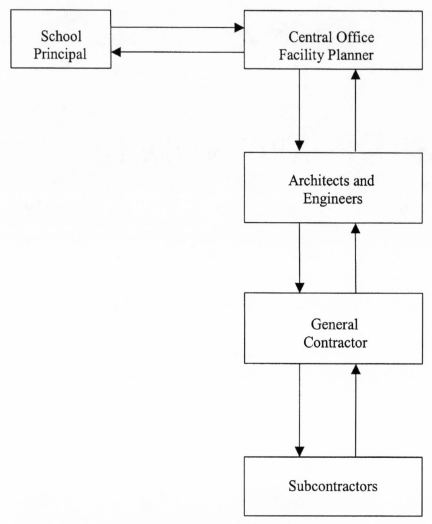

Figure I.1. Line of Communication in a School Renovation Project

APPENDIX J

PROJECT EVALUATION REPORT

Name of Project: _____

Date of Evaluation: _____ Weather Condition: _____

Temperature: _____

Attendants: _____

Materials and Equipment on Site: _____

_____

Materials and Equipment in Order: _____

_____

Workmen on Site: _____

_____

_____

Work Completed: _____

_____

_____

Work in Progress: _____

_____

_____

_____

Progress per Schedule: _____

Inspection by County/City Inspectors: _____

Items of Concern: _____

_____

Discussion: _____

_____

Notes: _____

_____

Evaluation Follow-ups: _____

_____

_____

Report submitted by: _____
Copied to: _____

_____

_____

REFERENCES

Abramson, P. (2002). School construction. *School Planning and Management, 41*(12), F-7.

Alexander, K., & Salmon, R. G. (1995). *Public school finance.* Needham Heights, MA: Allyn and Bacon.

Aller, G. (2002). Alternate project delivery methods: New ways to build and renovate school facilities. *School Business Affairs, 68*(11), 11-14.

Alvord, D. J. (1971). Relationships among pupil self-concept, attitude toward school, and achievement on selected science exercises for the National Assessment of Educational Progress. Unpublished doctoral dissertation, Iowa State University, Ames, Iowa.

Americans with Disabilities Act. (1990). Public Law No. 101-336, codified at 42 U.S.C. Sec. 12101.

Argon, J. (1997). Rising to new heights. *American School and University, 69*(9), 14-22.

Argon, J. (1998). Boom time. *American School and University, 70*(9), 20-31.

Argon, J. (2000). Through the roof. *American School and University, 72*(9), 30-45.

Argon, J. (2001). Building for the boom. *American School and University, 73*(9), 22-37.

Atwater, M., Gardner, C., & Wiggins, J. (1995). A study of urban middle school students with high and low attitudes toward science. *Journal of Research in Science Teaching, 32*, 665-677.

Baule, S. (1999). First steps in planning for facilities renovation. *Library Talk, 12*(5), 6-7.

Bennett, J. I. (1983). School renovation for a profit. *CEFP Journal, 21*(3), 22-24.

Berry, M. A. (2002). Healthy school environment and enhanced educational performance: The case of Charles Young Elementary School, Washington, DC. Retrieved April 9, 2003 from http://www.carpet-rug.com/pdf_word_docs/020112_Charles_Young.pdf.

Bowers, J., & Burkett, C. W. (1989). Effects of physical and school environment on students and faculty. *Educational Facility Planner, 27*(1)28–29.

Boyles, D. R. (1995). The corporate takeover of American schools. *Humanist, 55*(3), 20–24.

Bradley, W. S., & Protheroe, N. (2003). The principal's role in school construction and renovation. *Principal, 82*(5), 38–41.

Brent, B., & Cianca, M. (2001). Involving principals in school renovations: Benefit or burden? *Journal of Education Finance, 27*, 729–740.

Burrup, P. E., Brimley, V., & Garfield, R. R. (1996). *Financing education in a climate of change* (6th ed.). Needham Heights, MA: Allyn and Bacon.

Butterfield, E. (1999). Planning today for tomorrow's technology. *School Construction News, 2*(3), 16–17.

Carter, S. P. (2003). Safety and security by design. *School Planning and Management, 42*(5), 46–47.

Cash, C. S. (1993). Building condition and student achievement and behavior. Unpublished doctoral dissertation, Virginia Polytechnic Institute and State University, Blacksbury, VA.

Castaldi, B. (1982). *Educational facilities: Planning, modernization, and management* (2nd ed.). Boston: Allyn and Bacon.

Castaldi, B. (1994). *Educational facilities: Planning, modernization, and management* (4th ed.). Boston: Allyn and Bacon.

Cecil, D., & Boynton, R. (1999). A true community school. *School Planning and Management, 38*(4), 44–48.

Chan, T. C. (1979). The impact of school building age on the academic achievement of eighth grade pupils from the public schools in the state of Georgia. Ann Arbor, MI: University Microfilms International.

Chan, T. C. (1980). Initial costs vs. operational costs. A study of building improvement projects in fourteen schools in the school district of Greenville County, South Carolina. Greenville, SC: The School District of Greenville County (ERIC Document Reproduction Service No. ED 208 492).

Chan, T. C. (1980, June 3). Educational environment and student achievement. *Ming Pao Daily*, Hong Kong, p. 6.

Chan, T. C. (1982). A comparative study of pupil attitudes toward new and old school buildings. Greenville County School District, Greenville, SC (ERIC Document Reproduction Service No. ED 222 981).

Chan, T. C. (1983). The pros and cons of contractor financed approach to school construction. *CEFP Journal, 21*(6), 13.

Chan, T. C. (1988a). The aesthetic environment and student learning. *School Business Affairs, 54*(1), 26–27.

Chan, T. C. (1988b). Implementing an effective program for playground safety. *School Business Affairs, 54*(7), 44–46.

Chan, T. C. (1996a). Environmental impact on student learning. Valdosta State University, Valdosta, GA (ERIC document Reproduction Services No. EA 028 032).

Chan, T. C. (1996b, November). Physical environment and student safety in South Georgia schools. A paper presented at the annual meeting of the Georgia Educational Research Association, Atlanta, GA.

Chan, T. C. (1999). How to manage a new school building. *Principal*, November, 25–26.

Chan, T. C. (2000). Beyond the status quo: Creating a school maintenance program. *Principal Leadership, 1*(3), 64–67.

Chan, T. C. (2002). School building designs: Principles and challenges of the 21st Century. *School Business Affairs, 68*(4), 16–20.

Chan, T. C., Richardson, M. D., & Jording, C. (2001). Carpet in schools: Myth and reality. *School Business Affairs, 67*(6), 19–22.

Cianca, M., & Brent, B. O. (2002). Why districts should get principals involved in renovation projects. *School Business Affairs, 68*(8), 13–16.

Coffey, A. B. (1992). Revitalization of school facilities. Unpublished doctoral dissertation, East Tennessee State University, Johnson City, TN.

Coley, J. D. (1989). "Wolves at the schoolhouse door": Comments and impressions by a besieged business manager. *School Business Affairs, 55*(10), 10–13.

Coley, R., Cradler, J., & Engel, P. (1997). *Computers and classrooms*. Princeton, NJ: Educational Testing Service.

Connor, W. F. (1998). Six rules for school boards. *School Planning and Management, 37*(2), 61–62.

Cook, C. W., & Hunsaker, P. L. (2001). *Management and organizational behavior* (3rd ed.). New York: McGraw-Hill.

Cooze, J. (1995). Curbing the cost of school vandalism: Theoretical causes and preventive measures. *Education Canada, 35*(3), 38–41.

Corcoran, T. B., Walker, L. J., & White, J. L. (1998). *Working in urban schools*. Washington, DC: Institute for Educational Leadership.

Cotton, K., & Linik, J. R. (2000). Part-time class size reduction at Fall City Elementary. *The CEIC Review, 9*(2), 17.

Council of Chief State School Officers. (2000). *Standards for school leaders*. Washington, DC: Author. Retrieved February 2, 2003 from http://www.ccsso.org/standards.html.

Cramer, R. J. (1976). Some effects of school building renovation on pupil attitudes and behavior in selected junior high schools. Unpublished doctoral dissertation, University of Georgia, Athens, GA.

Cuban, L., Kirkpatrick, H., & Peck, C. (2001). High access and low use of technologies in high school classrooms: Explaining an apparent paradox. *American Educational Research Journal, 39*(4), 813–834.

Cunningham, C. (2002). Buildings that teach. *American School & University*, August, 164–167.

Davis, L., & Tyson, B. (2003). Strategic bonding. *American School Board Journal, 190*(7), 34–36.

Dawson, C. G. B., & Parker, D. R. (1998, November). A descriptive analysis of the perspectives of Neville High School's teachers regarding the school renovation. Paper presented at the annual meeting of the Mid-South Educational Research Association, New Orleans, LA.

De Haan, A. B. (2000). Renovations: Take the time to do it right. *School Planning and Management, 39*(3), 26, 28–32.

DeJong, W., & Glover, T. (2003). Planning requires cooperation. *School Planning and Management, 42*(4), 21.

DiBella, C. M., & Anderson, J. (2000). With apologies to Maria Shriver: 10 things you must know before starting a school construction or renovation project. *School Business Affairs, 66*(12), 30–33.

Drucker, P. F. (1999). Beyond the information revolution. *Atlantic Monthly, 284,* 47–57.

Duke, D. L., Griesdorn, J, Gillespie, M., & Tuttle, J. B. (1998). *Where our children learn matters: A report of the Virginia school facilities impact study.* Charlottesville VA: The Thomas Jefferson Center for Educational Design.

Dunklee, D. R., & Siberman, R. M. (1991). Healthy buildings keep employees out of bed and employers out of court. *School Business Affairs, 57*(12), 19–22.

Earthman, G. I. (1994). *School renovation handbook: Investing in education.* Lancaster, PA: Technomic Publishing.

Earthman, G. I. (1998). Renovations: Know the process, avoid the pitfalls. *High School Magazine, 5*(5), 31–35.

Earthman, G. I., Cash, C. S., & Van Berkum, D. (1996). Student achievement and behavior and school building condition. *The Journal of School Business Management, 8*(3), 26–37.

Egelson, P., & Harman, P. (2000). Ten years of small class size in Burke County, North Carolina. *The CEIC Review, 9*(2), 14.

Flanigan, J. L., Richardson, M. D., & Stollar, D. H. (1995). *Managing school indebtedness: A complete guide to school bonding* (2nd ed). Lancaster, PA: Technomic.

Fuller, S. K., & Petersen, S. R. (1996). *Life-cycle costing manual for the federal energy management program.* Gaithersburg, MD: National Institute of Standards and Technology, Building and Fire Research Laboratory.

Futral, K. K. (1993). The principal's role in school renovation. *Principal, 72*(3), 30–33.

Gardner, J. C. (1977). Life-cycle costing. *American School and University, 49*(12), 8, 10.

Geiger, P. E. (2001). School construction: Changing to meet the new trends in education. *School Business Affairs, 67*(12), 8–11.

General Accounting Office. (1995). *School facilities: Conditions of America's schools.* Washington, DC: Author.

Gisolfi, P. A. (1999). Reinventing schools. *Learning By Design, 8,* 12–15, 17.

Gooden, J. S., & Carlson, R. L. (1997). Achieving quality schools through technology change. In M. D. Richardson, R. L. Blackbourn, C. Ruhl-Smith & J. A. Haynes, *The pursuit of continuous improvement in educational organizations* (pp. 195–206). Lanham, MD: University Press of America.

Gottwalt, T. J. (2003). Preventive safety measures. *School Planning and Management, 42*(6), 26–28.

Hack, W. G., Candoli, I. C., & Ray, J. R. (1998). *School business administration: A planning approach* (6th ed). Needham Heights, MA: Allyn and Bacon.

Hansen, S. (1992). *Schoolhouse in the red: A guidebook for cutting our losses. Powerful recommendations for improving America's school facilities.* Arlington, VA: American Association of School Administrators.

Hartman, W. T. (1988). *School district budgeting.* Englewood Cliffs, NJ: Prentice Hall.

Hawkins, H. L., & Lilley, H. E. (1998). *Guide to school facility appraisal.* Scottsdale, AZ: Council of Educational Facility Planners, International.

Healthy Schools Network, Inc. (2000). *Guide to school renovation and construction: What you need to know to protect child and adult environmental health.* Albany, NY: Author.

Hebert, E. A. (1998). Design matters: How school environment affects children. *Educational Leadership*, September, p. 69–70.

Henry, C. (2000). Planning without anxiety. *School Planning and Management, 39*(10), 28–30.

Henry, R. A. (2001). Fixing it right. *American School & University*, September, 45–47.

Hill, F. (2000). Change is good . . . Wrong! *School Business Affairs, 66*(12), 41–43.

Hines, E. W. (1996). Building condition and student achievement and behavior. Unpublished doctoral dissertation, Virginia Polytechnic Institute and State University, Blacksburg, VA.

Hoffman, K. (2002). Designing with maintenance in mind. *School Planning and Management, 41*(12), 34–35.

Holt, T., & Kirby, J. R. (1997). Purchasing practices: Two education allies warn of the pitfalls of low-bid buying. *School Planning and Management, 36*(1), 28–30.

Honeyman, D. S. (1990). School facilities and state mechanisms that support school construction: A report from the fifty states. *Journal of Education Finance, 16*(2), 247–249.

Honeyman, D. S. (1994). Finances and problems of America's school buildings. *Clearing House, 68*(2), 95–97.

Honeyman, D. S. (1998a). *Financing school facilities.* Reston, VA: Association of School Business Officials International.

Honeyman, D. S. (1998b). The condition of America's schools. *School Business Affairs, 64*(1), 8–16.

Hoskens, J. P. (2003). School construction funding and policies. *School Planning and Management, 42*(7), 12.

Illuminating Engineering Society of North America. (2000). *Recommended practice on lighting for educational facilities.* New York: Author.

Jacobs, B. W. (1995). *Maintaining acceptable indoor air quality during the renovation of a school.* Baltimore, MD: Maryland Department of Education, School Facilities Branch (ERIC Document Reproduction Services: ED 415 661).

Jarvis, O. T., Gentry H. W., & Stephens, L. D. (1967). *Public school business administration and finance: Effective policies and practices.* West Nyack, NY: Parker.

Jones, E. (2002). School funding inequities: A statistical analysis examining the adequacy of funding for capital outlay in North Carolina schools. *Research for Educational Reform, 7*(1), 24–45.

Jozwiak, D. (1998). Preventive medicine. *American School and University, 70*(6), 16, 18.

Kalinger, P. (1998). The benefits of preventive roof maintenance. *School Planning and Management, 37*(6), 44–47.

Kendler, P. B. (2003). Hard hat required. *District Administration, 39*(12), 26–29.

Kennedy, M. (1999). Closing doors. *American School and University, 71*(11), 16–22.

Kennedy, M. (2000). A well-grounded plan. *American School and University, 72*(10), 30–34.

Kirk, S. J., & Dell'Isola, A. J. (1995). *Life-cycle costing for design professionals.* New York: McGraw Hill.

Kissing, S. (1998). Designing schoolhouses of quality. *School Planning and Management, 37*(6), 21–26.

Kosar, J. E. (2002). Cultivating dialogue before building. *School Administrator, 59*(6), 28–30.

Kowalski, T. J. (1989). *Planning and managing school facilities.* New York, NY: Praeger.

Kowalski, T. (2002). *Planning and managing school facilities* (2nd ed.). Westport, CT: Bergain & Garvey.

Lackney, J. (1999, January). Reading a school building like a book: The influence of the physical school setting on learning and literacy. A paper presented at the PREPS Conference in Jackson, MS.

Lane, K. E., & Richardson, M. D. (1993). Opening a new school with effectiveness: Three steps to success. *PEB Exchange, 19*, 5–7.

Lane, K. E., Richardson, M. D., & VanBerkum, D. W. (1993). Technology in the classroom: Does it impact learning? *Journal of School Business Management, 5*(1), 21–25.

Leibowitz, J. (2001). Making the grade. *Facilities Design and Management, 20*(5), 42–47.

Lemasters, L. K. (1997). A synthesis of studies pertaining to facilities, student achievement, and student behavior. Unpublished doctoral dissertation, Virginia Polytechnic Institute and State University, Blacksburg, VA.

Linn, H. (1952). Modernizing school buildings. *American School and University, 24*, 401.

Macclay, W. R., & Earthman, G. I. (1992). Post-occupancy evaluation of Standley Lake High School. *Educational Facility Planner, 30*(3), 7.

MacKenzie, D. G. (1989). *Planning educational facilities.* New York: University Press of America.

MacKenzie, D. G., & Phillips, P. (1991). Determining school district renovation/remodeling/repair needs. *Educational Facility Planner, 29*(4), 9–11.

Maxwell, L. E. (1999). School building renovation an student performance: One district's experience (ERIC Document Reproduction Services: ED 443 272).

McGuffey, C. W. (1972). *Pupil attitudes toward existing school as compared to new fully carpeted, air-conditioned schools.* Athens, GA: University of Georgia.

McRobbie, J. (1997). Class-size reduction: A one-year status check. *Thrust for Educational Leadership, 27*(1), 6–10.

Means, B., & Olson, K. (1995). *Restructuring schools with technology.* Menlo Park, CA: SRI International.

Mearig, T., Coffee, N., & Morgan, M. (1999). *Life-cycle cost analysis handbook.* Juneau, AK: State of Alaska, Department of Education and Early Development.

Moore, D. (2002). We can't put it off forever. *School Planning and Management, 41*(5), 44–47.

Moussatche, H., Languell-Urquhart, J., & Woodson, C. (2000). Life-cycle costs in education: Operations & maintenance considered. *Facilities Design and Management, 19*(9), 20–22.

National Center for Education Statistics and Association of School Business Officials International. (2003). *Planning guide for maintaining school facilities.* Washington, DC: Author.

National Clearinghouse for Educational Facilities—News. (2003, April 25). School maintenance and operations spending down. Retrieved May 9, 2003, from http://www.edfacilities.org/ne/news.html.

National Education Association. (2002). *School modernization.* Washington, DC: Author. Retrieved January 23, 2003, from http://www.nea.org/mordernization.

National Policy Board for Educational Administration for the Educational Leadership Constituent Council. (1995). NCATE program standards. Alexandria, VA: Author.

Nixon, C. W. (1998). Today's schools, tomorrow's classrooms. *School Planning and Management*, *37*(11), 26, 28, 30.

Peters, R., & Smith, M. (1998). New schools from old space. *Educational Facility Planner*, *34*(3), 14–16.

Pittillo, R. A. (1993). Maintenance: The unavoidable. *Educational Facility Planner*, *31*(4), 20–21.

Proshansky, H. M., Ittelson, W. H., & Rivlin, L. G. (1970). *Environmental psychology: Man and his physical setting*. New York: Holt, Rinehart & Winston.

Rabenaldt, C., & Velz, E. (1999). Lessons learned in school design and construction. *School Planning and Management*, *38*(5), 39–44.

Reicher, D. (2000). Nature's design rules. *Learning By Design*, *9*, 16–18.

Responsible Industry for a Sound Environment. (1999). A bug's life. *American School and University*, February, 39–41.

Richardson, M. D., Chan, T. C., & Lane, K. E. (2000). Planning for technology: Issues and concepts. *Educational Planning*, *12*(4), 67–75.

Richardson, M. D., Gentry, L. R., & Lane, K. E. (1994). Teacher science lab liability: Protect yourself when administrators don't!!! *Journal of Chemical Education*, *71*(8), 689–690.

Richardson, M. D., & Lane, K. E. (1997). Learning as continuous improvement in educational organizations. In M. D. Richardson, R. L. Blackbourn, C. Ruhl-Smith & J. A. Haynes, *The pursuit of continuous improvement in educational organizations* (pp. 55–74). Lanham, MD: University Press of America.

Richardson, M. D., Short, P. M., & Lane, K. E. (1996). Synergistic planning: A useful tool in school-based management. *Educational Planning*, *10*(3), 13–20.

Ridler, G. E., & Shockley, R. J. (1989). *School administrator's budget handbook*. Englewood Cliffs, NJ: Prentice Hall.

Ritter, G. W., & Lucas C. J. (2003). Estimations of readiness for NCLB: Lessons learned from state education officials. *Education Next*, *3*(4). Retrieved November 24, 2003, from www. educationnext.org.

Rittner-Heir, R. (2003). Schools and economic development. *School Planning and Management*, *42*(4), 16–18.

Rivera-Batiz, F. L., & Marti, L. (1995). *A school system at risk: A study of the consequences of overcrowding in New York City Public Schools*. New York, NY: Columbia University, Institute for Urban and Minority Education.

Rochefort, M., & Gosch, J. (2001). Avoiding construction snafus. *American School and University*, *74*(3), 347–348.

Roundtree, M. L. (1997). The state-initiated class-size reduction program: A preliminary study of the initial district response. Doctoral dissertation, University of Southern California, Los Angeles, CA.

Ryland, J. (2003). Fads, fancies and fantasies: An educator's perspective on current educational facility issues. *School Planning and Management*, *42*(6), 16–20.

Salmon, R. G., & Thomas, S. B. (1981). Financing public school facilities in the '80s. *Journal of Education Finance*, *7*(1), 11–109.

Sanoff, H. (2002). *Schools designed with community participation*. Washington, DC: National Clearinghouse for Educational Facilities.

Schaefer, J. W. (1967). *What is operations? A handbook for school business officials.* Chicago, IL: The Research Corporation of the Association of School Business Officials.

Schneider, M. (2002). Do school facilities affect academic outcomes? Retrieved April 9, 2003, from http://www.edfacilities.org/pubs/outcomes.pdf.

Schofield, J. (1995). *Computers and classroom culture.* London: Cambridge University Press.

Sharp, W. L. (1992). Preventative maintenance: Toward an expanded concept. *Educational Facility Planner, 30*(4), 11–12.

Shaw, R. (2000). Clean and safe. *American School and University,* September, 42–43.

Shideler, L. (2001). A clean school is a healthy school. *American School and University,* May, 52–56.

Simko, E. A. (1987). Proactive maintenance saves money. *American School and University, 59*(7), 31–38.

Smith, J. M., & Ruhl-Smith, C. (2003, February). Corporate solutions to public school effectiveness: An examination and comparison of ineffective business practices applied to school leadership. A paper presented to the American Association of School Administrators Annual Conference in New Orleans, LA.

Spitz, K. (2001). Designing campus landscapes for preventive maintenance. *College Planning and Management, 4*(3), 52–53.

Stapleton, D. B. (2001). Differences in school climate between old and new buildings: Perceptions of parents, staff and students. Unpublished doctoral dissertation, Georgia Southern University.

Stevenson, K. R., & Terril, T. (1988). Avoid dissention by involving citizens in facilities planning. *American School Board Journal, 175*(11), 30.

Stollar, D. H. (1967). *Managing school indebtedness.* Danville, IL: Interstate Publishers.

Strickland, J. S., & Chan, T. C. (2002). Curbside critique: A technique to maintain a positive school yard image. *School Business Affairs, 68*(5), 24–27.

Sturgeon, J. (2000). This space occupied. *School Planning and Management, 39*(8), 37–40.

Swanson, A. D., & King, R. A. (1997). *School finance: Its economics and politics.* White Plains, NY: Longman.

Tanner, C. K., & Morris, R. F. (2002). School physical environment and teacher and student morale: Is there a connection? *School Business Affairs, 68*(10), 4–8.

Thompson, D. C., & Wood, R. C. (1998). *Money and school.* Larchmont, NY: Eye On Education.

Thompson, D. C., & Wood, R. C. (2001). *Money and school* (2nd ed.) Larchmont, NY: Eye On Education.

Thompson, D. C., Wood, R. C., & Honeyman, D. S. (1994). *Fiscal leadership for schools: Concepts and practices.* White Plains, NY: Longman.

Thompson, T. (1991). People make the difference in school playground safety. *The Executive Educator, 13*(8), 28–29.

Tressler, P. M. (1997). Initial implementation of the class size reduction program in Orange County: A survey of the issues (California). Doctoral dissertation, University of Southern California, Los Angeles, CA.

U.S. Consumer Product Safety Commission. (1981). *Handbook for public playground equipment and surfacing.* Washington, DC: Author.

U.S. Consumer Product Safety Commission. (1993). *Handbook for public playground safety.* Washington, DC: Author.

U.S. Department of Education. (1999). *Modernizing America's schools for the 21st century.* Washington, DC: Author.

U.S. Department of Education. (2000). *Impact of inadequate school facilities on student learning.* Washington, DC: Author.

U.S. Department of Education. (2000, April). *Schools as centers of community: A citizens' guide for planning and design.* Retrieved January 23, 2003, from http://www.ed.gov/inits.construction/commguide.pdf.

U.S. Department of Education, National Center for Education Statistics, & National Forum on Education Statistics. (2003). *Planning guide for maintaining school facilities.* Washington, DC: Authors.

U.S. General Accounting Office. (1996). *School facilities: America's schools report differing conditions.* Washington, DC: Author.

Vasfaret, G. (2002). Preventive maintenance: Fighting time and elements. *School Business Affairs, 68*(11), 6–10.

Westerkamp, T. (2000). Preventing plumbing problems. *Maintenance Solutions Online,* December, 2000. Retrieved March 6, 2003, from http://www.facilitiesnet.com/ms/Dec00/dec00interior.shtml.

White, E. T. (1992). Post-occupancy: A new component in the building delivery process. *Educational facility Planner, 30*(3), 4.

Williams, A. J. (2002). Superintendents' perceptions regarding the impact of class size reduction on school facility planning in Georgia. Doctoral dissertation, Georgia Southern University, Statesboro, GA.

Wood, R. C. (1986). *Principles of school business management.* Reston, VA: Association of School Business Officials International.

Wright, D. (1996). Apples to apples. *School Planning and Management, 35*(7), 19–21.

INDEX

ABOUT THE AUTHORS

Tak Cheung Chan, professor of educational leadership and coordinator of doctoral program, Kennesaw State University, Georgia, received his doctor of education degree from the University of Georgia. He also holds a bachelor of arts degree from the University of Hong Kong and a master of education degree from Clemson University. Dr. Chan was a language and social studies teacher, assistant school principal, and school principal in Hong Kong, where he was born and raised. His American public school experiences include serving as project administrator, facility manager, and educational planning director at three major school districts in Georgia and South Carolina. His career in higher education began as an adjunct professor at Kennesaw State University and an assistant professor of educational leadership at Valdosta State University in Georgia. He then served as an associate professor of educational leadership at Georgia Southern University. With more than 30 years in education, Dr. Chan shares his unique experiences in professional publications and conferences. He is a prolific writer with a keen interest in school facilities and business management. He is also a frequent presenter at educational conferences at state, national, and international levels. His research focuses on educational planning, international education, and current educational leadership issues.

Michael D. Richardson is professor of educational administration at Georgia Southern University. Dr. Richardson came to Georgia Southern University from Clemson University, where he served as coordinator for doctoral programs and acting associate dean for research in the College of Education. Prior to Clemson, Dr. Richardson was at Western Kentucky University, where he served as coordinator of educational administration, research, and foundations. He completed a B.S. and M.A. in education at Tennessee Technological University and was awarded the Ed.D. from the University of Tennessee. Dr. Richardson serves as founding editor of the *Journal of School Leadership*, an internationally refereed journal of educational administration. He has served on editorial boards for more than 20 journals. In addition, he has authored or edited 14 books, published more than 95 articles in professional journals, and made more than 200 presentations to national and international professional organizations. Dr. Richardson served as a secondary and elementary principal, personnel director, director of special projects, coordinator of federal programs, and assistant superintendent before entering academe.

Dr. Richardson has researched and written extensively concerning organizational theory and the principalship. He has received awards for teaching and for service to the profession.